Pulp & Paper Technology

Technology, Testing & Applications

Pulp & Paper Technology

Technology, Testing & Applications

K.P. RAO
Manipal Institute of Technology,
Manipal

CBS PUBLISHERS & DISTRIBUTORS
NEW DELHI • BANGALORE

ISBN : 81-239-1004-5

First Edition: 2003

Reprint : 2007

Production Director : Vinod K. Jain

Published by :
Satish Kumar Jain for CBS Publishers & Distributors,
4596/1-A, 11 Darya Ganj, New Delhi - 110 002 (India)
E-mail : cbspubs@del3.vsnl.net.in
Website : http://www.cbspd.com

Branch Office :
Seema House, 2975, 17th Cross, K.R. Road,
Bansankari 2nd Stage, Bangalore - 560070
Fax : 080-6771680 • E-mail : cbsbng@vsnl.net

Printed at :
Chaman Enterprises, New Delhi - 110 002

Preface

During my teaching the subject of paper, I had prepared a lot of handwritten teacher's notes. As is expected, the handwriting was quite unreadable; hence, I decided to have the text printed in a computer. At the end of the year I found that the matter was very different than any of the books that existed on the subject.

My students who had a look at my notes said that it would be great if my notes could be published as a book with the sketches at the appropriate places.

Meanwhile one of our friends Dr. Rajagopal Shenoy was writing a book on surgery and recasting the book was being done by us here. This was the kind of fillip that was needed to start on the job. I am very thankful to Dr. Shenoy for getting us started on the job.

The help that I got from my colleagues in the department was good. Mr. Ganesh, Mr. Jayaram and Mr. Mukund have been a great help in bringing out this book. I am also thankful to my students, Mr. Sudhir Santosh for the figures he provided and to Ms Mumtaz Bolwar for going through the manuscript.

Mr. B.S. Bairi, vice principal (academic) at Manipal Institute of Technology and Dean (Faculty of Engineering), Mangalore University, encouraged me to bring out this book. I am grateful to him for his good words that kept me going.

Manipal   K.P. Rao

Contents

SECTION 2

Properties and Testing of Paper 73-116

SECTION 3

Applications 117-140

Introduction

1.1 PAPER IN OUR SOCIETY

We live in an information age. The operations performed on information are acquisition, documentation and transfer. Hand written or printed books have been the most important way of information archival and transfer in time. The process of writing and printing involves creating a visual contrast on a substrate. This is introduced with a deliberate variation of colour or shape on the substrate. Printing involves the process of reproduction of such contrast by transferring the image from a printable surface to a substrate in multiple copies.

The earliest attempt to write possibly was a line drawn on sand to indicate direction. With some training it would have been possible to draw pictures and some ideas could be conveyed. A more complex invention of mankind followed when a visual representation of the spoken syllable of language was attempted in terms of writing. Writing defies time and occupies space unlike the spoken word which exists entirely in time. The process of writing involves encoding the spoken word or a concept into a form. When the form represents a concept or a thing it is known as the ideogram. Attempts to communicate with the help of pictures was a common factor of many ancient cultures from all around the world. Later experimentations led to the development of phonetic writing where a sound value is assigned to each character rather than an idea. Combination of these phonograms produces a complete word which may be built into sentences.

All writing and most of printing is done on paper, though printing can be done on many other substrates. Paper has some amazing properties that makes it the most popular medium for information dissemination. It is light and thin, hence can be built into very portable high information density packages such as books. It is strong enough to stand the mechanical marking operations. It can be printed and made into attractive publications. It is not expensive and has good storage and handling properties. It is semi-permanent in that whatever that is written on cannot be easily erased. Materials other than paper turn important only in packaging applications.

1.2 A BRIEF HISTORY

In India the layered bark of the *Bhoorja* tree was used for writing. *Bhoorja* tree grows in the Himalayan ranges and is nature's own paper making plant. The material is brownish in colour, thin, very flexible and easily marked with inks. It is however not available in large sizes and may contain shives or knots in its body. The leaves of the palm tree treated properly provided the engravable writing material. Cloth coated with clay and gum was used as a writing medium. Cloth formed the base for a flexible, black to gray coloured, highly polished coated surface that could be bound into books or used as single sheets.

Paper consists of a web or mat of inter meshed cellulosic fibers. The mat is formed when a very dilute aqueous suspension of the separated fibers is dried by draining off the water leaving the fibers to form a settled layer. Many additives are contained in paper and often a variety of surface treatments are done to make paper suitable for the applications it is made for. The thickness of the material produced and the finish decide if it is board or paper.

The term paper is derived from the word papyrus. This was a paper like material which was used for writing by the Egyptians more than 3000 years ago. Papyrus is a plant (*Cyperus papyrus*) that grows on the banks of the Nile. The inside of the plant was cut into thin strips and they were glued together to form a thin, semi flexible, translucent to opaque, flat material. The papyrus writings have stood the test of time for the last 3000 years.

Historically the Chinese invented the modern paper (Tssi Lun in 105 A.D.), and made it using pulp. The pulp was made from shredded cloth fibers and bamboo fibers. These materials were heated in milk of lime. The resulting liquid was poured into a frame covered with loosely woven cloth. The water drained through the cloth leaving a mat like material which was dried in the sun to produce paper. The laid paper was also made by Tssi by immersing the screen in the pulp and gently raising the screen to form the paper. From China the process of paper making reached Europe via Constantinople. In the medieval period there was a huge shortage of rag as most of the paper making used rag as its basic raw material. Alternate sources were sought after and using fibers directly from plants was considered.

1.3 INDUSTRIAL PRODUCTION OF PAPER

The hand process was used for paper making till 1800 in Europe. The first automated process was developed in the paper mill of M. Didot in France by Nicolas Louis Robert. The process involved pouring the fiber suspension on a continuously moving wire belt. The water would drain off leaving a continuous length of paper, which was further dried by pressing between felt covered rollers. The machine's development ended with the French revolution. John Gimble an associate of the Didot family joined the Fourdrinier brothers in England to put Robert's ideas into a new machine. By 1807 they had designed and built the first fully automated continuous paper-making machine. The stuff—a previously tested suspension of woody material—was deposited from a special box onto a continuously moving copper alloy belt. The belt was 54 inches wide and 31 feet long. The belt moved both forward and side to side. The machine with all its other parts was more than 100 feet long. It produced paper at the rate of 600 feet an hour. The paper produced as a continuous web was cut by hand and was allowed to air dry. In 1839 Robert Ransen produced the first on line dryer and the paper was wound on a large roll. These machines were copied by many all over the world. Almost all the paper made today is made on the fourdrinier paper-making machine or its variations. Only a negligible little is hand made.

In 1839 chemists worked on the identification of cellulose. With the development of cellulose chemistry, paper formation and technology was better understood. The scientific study of paper started after 1870 when pulping was studied in greater detail by chemists and chemical engineers.

As of today wood is the main starting material for making most of the paper.

Paper making consists of the following steps:

1. **Making of cellulose pulp:** This is done from various sources of fiber mainly wood and other plant material.

2. **Preparation of the stock:** This involves the mechanical treatment of the fibers such as beating and refining. The purpose is to render desired properties to the finished product.

Pulps from many sources may be blended together, non-fibrous materials may be added and the dilute suspension is ready for the paper making.

3. **Paper machine operations:** The paper is made by two processes. It may be made by the hand process or made on automatic paper-making machines. In the hand process pulp is drained free of water by pouring onto a fine mesh or lifting off a deckle. This produces the laid paper. While making paper by the machine process, water is removed initially by draining off, then by suction and after that by application of heat and pressure. The paper comes out in a continuous web.

The paper may pass through stages of calendering, coating, slitting and cutting to the desired size after which it is repacked in appropriate sizes and transported to the destination.

In a paper plant the sequence of processes involved in paper making may be summarized as follows:

The process starts with the debarking where the bark is removed from the cut trees. The wood is cut into chips in a chipping machine. The chips so produced are sieved and rechipped if necessary to produce chips between 3 and 25 mm length. Saw dust is also produced at this time. Chipping machines have capacities in the order of 25 tonnes per hour. The chips are cooked with sodium hydroxide and sodium sulfate in digesters with a capacity of 80 cubic meters. Lignin in wood is converted into soluble compounds when cooked for about 5 hours. The chemicals are recovered to the best possible extent. The dark brown pulp is washed and bleached using chlorine, peroxides and hypochlorites. The pulp so obtained may be mixed with soft wood pulp to improve the quality of paper. The pulp at high dilution is drained on the paper machine to produce paper. The material is dried by pressing and then passed through dryers. Surface sizing and coating may be done to improve the qualities of paper. The paper is rewound, cut and slit into appropriate sizes.

Paper has more uses than just as a writing medium. A thicker variation of paper—board which is made by very similar processes is used in packaging. The purpose of packaging is to protect goods from damage and deformation in transportation from the manufacturer to the user. Packaging can also be decorative when printing may be done on the packages to attract the customer.

Paper is also used in electrical applications. This is because the paper is a good insulator which is very flexible, can be made into sheets that are reasonably thin and uniform and can also be pore free and consistent. The paper may be immersed in oils to improve the electrical properties further.

1.4 FUTURE OF PAPER

There has been some talk of a paperless society and a paperless office. There were many predictions to say that paper will totally disappear from all walks of life. The happenings have been quite to the contrary. The reasons for this are also quite obvious.

- Paper is by far the cheapest material that can cover the largest area in its given thickness. Any other material of an equal area will be orders costlier.
- Some of the disadvantages of paper turn out to be long-term advantages.
- Paper is not very stable and hence biodegradable unlike plastics.
- Because of the imperfect formation it is not as transparent as it could be, hence can be printed on both sides.
- It has high porosity which allows a quicker drying of ink than pore-free plastics.
- It is not rigid hence can be easily folded unlike metals.

- It is easy to make semi-permanent documents and it is also amenable to many modes of security.
- Alternates to printing on paper are very expensive and not portable.

All this and more gives paper a very unique place as a medium of communication.

Paper is not considered environment-friendly and increase in use is supposed to lead to deforestation. A balanced recycling of fiber and aforestation should make this allegation untrue.

SECTION

One

Technology

Paper Making

2.1 SOURCES OF FIBER

Paper is made from fibers. The fibers suspended in water is known as pulp. The procedures for obtaining pulp are different for different fiber sources.

Some of the sources of fiber are available only in some seasons while some others such as wood from forests is available at all times.

1. Wood fibers are fibers obtained from coniferous and deciduous woods. Most of the world's requirement of paper is met by paper from wood. There are three sources of wood.

   • Planned and cultivated forests which may be privately owned or owned by paper companies.

   • Naturally grown forests owned by Governments and private individuals cleared for farming or industry.

   • By-products of the wood-based industries that is available as chips and saw dust.

The properties of wood depend entirely on the species and the climatic conditions in which it is grown. Some of them are soft and some are hard.

2. Grass fibers such as straw, bamboo and bagasse.
3. Bast fibers such as kenaf, hemp, jute and linen
4. Leaf fibers such as esparto, sisal, pineapple are non-wood sources of fiber. They are grown mostly in vast farms. Bagasse is an important by-product of the sugar industry whereas the others may be grown specifically for the purpose of paper manufacture. Most of these sources contain chlorophyll, pith and other such material which are not useful for making paper.
5. Seed hairs such as cotton etc. are sources either directly as cotton or as pieces of cloth and thread. They are very expensive sources of fiber but are added to obtain some very special properties to the paper produced.

There are two more sources of fiber both of which depend essentially on the above basic sources. They are known as the secondary sources of fiber.

6. Rags and textile waste   Basically cotton and linen. Pure fibers with impurities.
7. Waste paper   Mixed woody fibers with lot of external impurities.

The fibers obtained from these sources differ in length, quality and other characteristics. The non-fibrous material is also very different in each case. The process of separation and cleaning of the fibers hence is very different for each source.

A comparison of lengths of fibers for some of the fibers of interest in paper-making is given in Table 2.1.

These variations in physical dimensions affect the performance of the product drastically. The way in which the fibers are packed into the body of the base material is also very different.

Table 2.1

Fiber	Length in mm	Width in mm
Cotton	10-50	0.025
Coniferous woods	4	0.025
Deciduous woods	1.5	0.030
Esparto	1.5	0.013
Straw	1.5	0.015

2.1.1 Wood as a source for pulp

Wood can be broadly classified into two categories i.e. soft or coniferous wood and the hard or deciduous wood. Trees that naturally grow in the colder regions of the world, such as fir, spruce, pine etc. fall into the softwood variety. Tropical and rain forest wood and the tress such as oak, cherry etc. are deciduous woods. The structure and constitution of woods are entirely different and so also the quality of the cellulosic fiber and hence the quality of the paper manufactured.

Plants grow with the help of sap (salts from the soil dissolved in water) which travels up the roots to the branches and leaves. Carbohydrates are formed by photosynthesis at the green pigment medium of the leaves in the presence of carbon dioxide from air and water through a series of complex chemical reactions. The carbohydrates so formed are carried to the parts of the plant that grow.

The annual growth in the trunk is marked by a thin layer immediately below the bark known as the cambium. The annual growth rings so formed in concentric layers are seen when the trunk is cut. The cambium grows by repeated division of cells which consists largely of long thin tubular cells known commonly as fibers.

Cell walls of wood are composed of cellulose and hemicellulose bound together by lignin into fibers. Cellulose has a chemical structure very similar to starch or sugar. The empirical formula of cellulose has been determined as $(C_6H_{10}O_5)_x$. Cellulose consists mainly of D-glucose units which can be hydrolyzed with nitric acid. The value of x may vary between 2000 in wood to about 500 in cotton. A single cellulose molecule in wood pulp is about 0.75 nm wide and between 800 nm and 5000 nm long. Lignin is a highly oxygenated aromatic polymer with a repeating phenyl propane skeleton. Chemically the fibers contain about 50% Carbon, 43.4% Oxygen, 6% Hydrogen, 0.1% Nitrogen and 0.5% Ash. On this matrix are deposited low molecular weight materials known as extractives. Plant cells, unlike the animal cells, have true cell walls containing polysaccharides as the major structural material. Cellulose is the major constituent of these cell walls. It exists in the cell wall as long thread like fibers— microfibrils—which are aggregated chains of cellulose. These molecules lie side by side along the fiber axis and together form the microfibrils which open out from the fiber due to beating. These fine threads are about 10 nm wide and of indefinite length. There are both crystalline and amorphous regions within these microfibrils. According to one of the studies, the cross-section of the microfibril is 3.5 nm × 3.5 nm and contains about 36 cellulose chains. The maturity of wood decides the way these microfibrils are embedded in lignin to form wood. Soft wood of less than 12 years of age, known as juvenile wood, has shorter fibrils. Paper made from them is likely to be poor in terms of tearing strength and may be better in terms of burst strength and fold resistance.

More than 90% of the volume of softwoods are tracheids. Tracheids are nothing but elongated and lignified single cells. The two end walls are tapered and overlap with other nearby tracheids. This is how they build mechanical strength and support the plant. When fully grown they are dead and have empty lumens. The edges of the tracheids have pits that transfer sap and water. Water hence passes from roots to the branches and leaves by this transfer process. Tracheids average 3 to 5 mm in length and width averaging 0.025 mm. Ray cells and epithelical cells are also noticed in softwood tracheids.

There are 3 kinds of hardwood fibers. Tracheids, fiber tracheids and libriform fibers. The last is most common and has dimensions as follows: Length 0.3 to 3 mm, Width 0.015 mm. Ray cells occur in large quantities in hardwood.

The wood fiber exists in four distinct layers surrounding the lumen—a central cavity. The outer layer—primary wall—has lot of lignin and a random arrangement of fibrils. In the inner layers S1, S2 and S3 the fibrils lie in a greater order. The S2 layer is the thickest.

The process of pulping is separation of the fibers from the unwanted material.

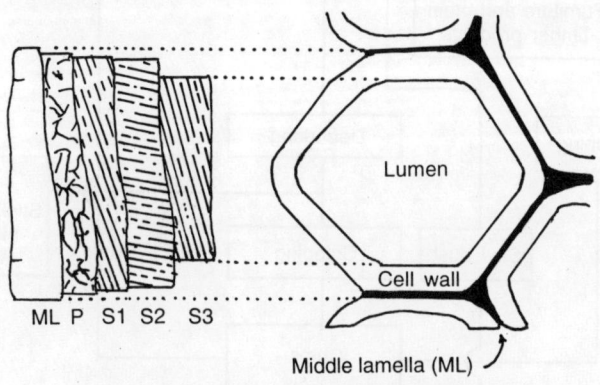

Fig. 2.1. The structure of wood fiber.

To the pulp maker and paper maker, separation of lignin from woody tissues is a major problem. Pulping could be understood as the technique of lignin removal. If lignin did not exist application of strong alkaline or acidic reagents would not have been necessary.

Lignin in ground wood pulp gives inferior optical properties and sometimes poor mechanical properties. Colour of lignin is caused by the presence of orthoquinones, catechols, free radicals and sometimes metal complexes.

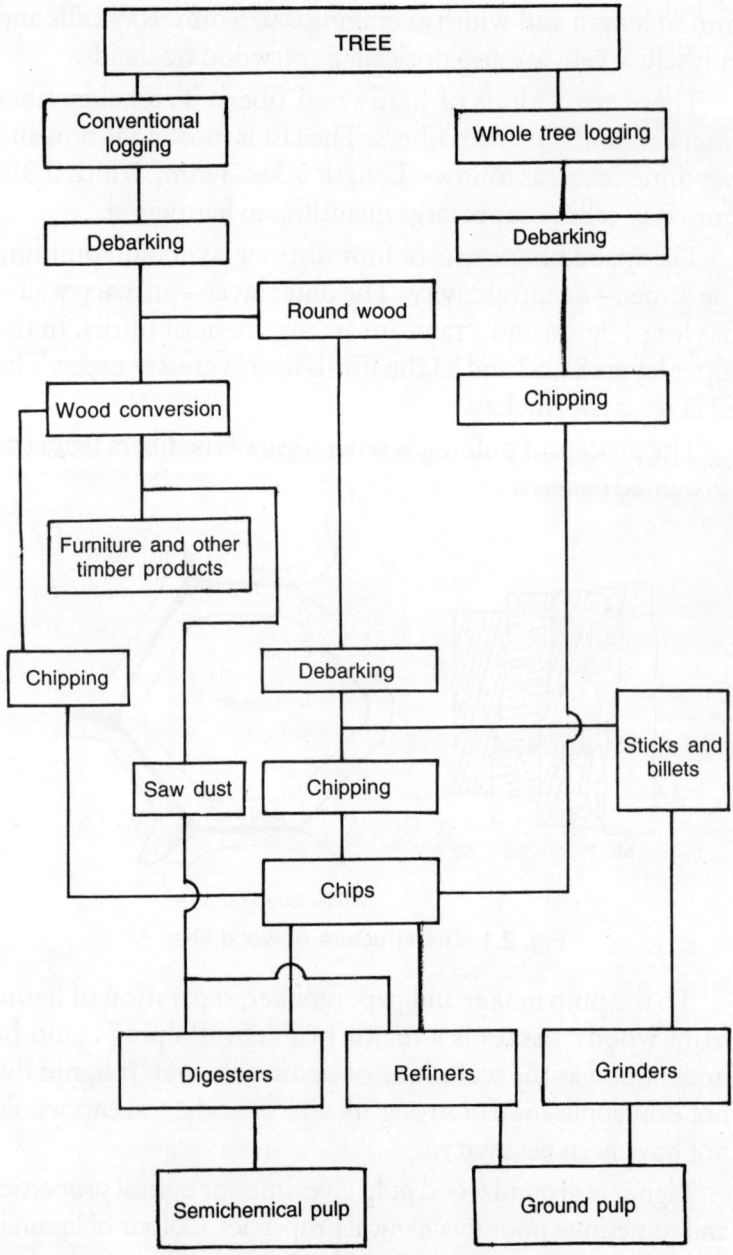

Fig. 2.2. Flowchart of pulping wood.

Pulpwood is debarked. The barks may have some useful chemicals, glues and other compounds which may be separated from unwanted materials in separate processes, otherwise it may be used as a fuel. Usually the barks are only contaminants to the pulp. They may. be used only as a fuel in some establishments. In some cases the whole tree and sometimes sized logs are loaded into the debarking units. In the tumbler debarker large logs of wood enter tumbling equipment from one side and are forced through the other. In the rotating drums pounding and abrasive action of the logs dislodges and tears off the bark which can easily be washed away. High pressure water jets are used in some cases to rip the bark off. After debarking the logs and sticks go either into grinding units directly or go into the chipping units.

Ground wood pulp (Stone Ground Wood Pulp): This does not use any chemicals and gained commercial importance in 1843. The process is to use large grinding stones continuously sprayed with water and the logs perpendicularly forced onto the stones. Friction breaks the wood down to its components. The temperature may reach 150 degrees in the grinding zone. The fibers are torn from the wood and carried to the pulp pit. All the lignin is carried as it is to the pulp. The yield is the largest of all known processes. Grinding involves the use of huge amounts of cheap electrical power and has restricted use in areas where cheap hydropower is available. The pulp so made is passed through knotter screens having 6 mm to 3 mm holes and then through primary and secondary screens. Centrifugal separators or cyclone cleaners (vortex cleaners) may also be used which remove besides the shives or larger pieces of wood any grinder grit. The tailings and rejects may be passed onto refiners.

There are two independent processes that happen in grinding. Wood fibers and fiber bundles are torn from the wood in the first stage and they are regrouped to a smaller size in the grinding operation.

The goal of wood grounding process is to get pulp that could be used directly as the furnish of the newsprint paper—that is to get paper of acceptable quality in terms of tearing strength, opacity, brightness, porosity and printability directly. This is not achievable most of the time.

Grinding of wood in Europe is usually done after preheating or cooking the wood. The wood may be boiled in water or may be heated at 110-125 degrees under pressure for 5 to 10 hours before grinding. The grinding time and power requirement is drastically reduced. The resulting pulp is generally coloured. The grinding time is reduced by treating wood with chemicals. A mixture of 6 parts of sodium sulphite and 1 part sodium bicarbonate is used in cooking wood before grinding. This is hence known as the semi-chemical pulping process. The separated fibers are uniform in the chemiground wood pulp. The brightness of the pulp is low and even bleaching does not help.

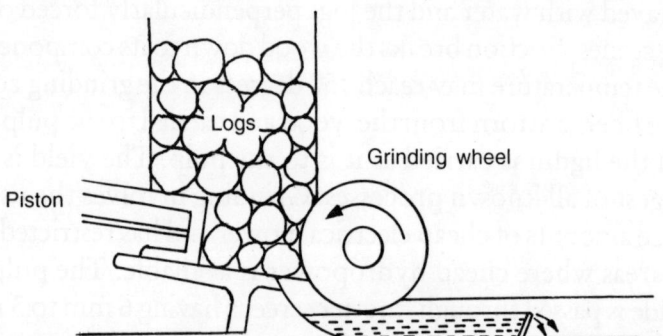

Fig. 2.3. Grinding wood.

When chips are formed from the debarked wood, multiple knife disk type chippers may be used. A heavy steel disk with knives protruding from its surface rotates at a high speed. The logs are fed with their ends into the path of the rotating knives. The cutting and shearing action slices the wood into chips. The chips are screened for a uniform size. They are rechipped if the pieces are too large. The chips may be stored in large heaps often for further processing. The transport of chips is done on conveyer belts or by blowing air.

The chips may be converted into pulp by many processes.

2.1.1.1 *Refiner Mechanical Pulping*

Refiners are used on chips and the process is most conveniently used on a variety of materials. The product unlike ground wood pulp is very consistent in quality. Disk refiner plate mechanisms are the most commonly used. The shredding takes place within the bars as the plates are at an angle. Certain amount of screening happens within the refiner itself. The refiner area usually runs hot and the heat generated by friction softens wood. Most of the refining takes place in the fine bar section. The refiners may be single stage or multiple stages.

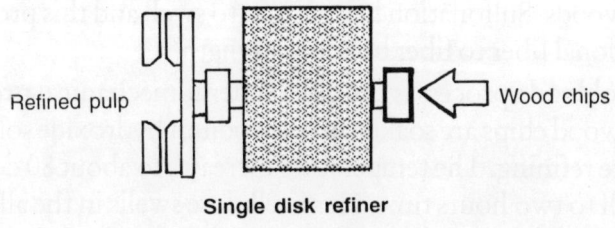

Single disk refiner

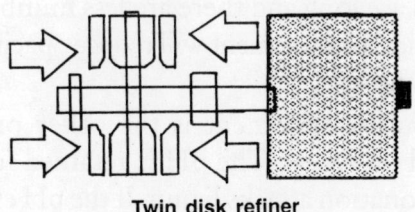

Twin disk refiner

Fig. 2.4. Refiner mechanisms.

2.1.1.2 *Thermo-Mechanical Pulping*

This involves softening of the lignin by application of steam under pressure on the chips. Steam at temperatures in excess of 165°C is applied on the refiners in closely coupled systems. The lignin is softened to such an extent that the fibers are separated in almost undamaged condition.

The equipment used may be single disk, twin disk or counter rotating disk refiners.

2.1.1.3 Chemi-Mechanical Pulping

Thermo-mechanical or mechanical pulping produces acceptable quality pulp from softwood. The ray cells of hardwood tend to affect the quality of the paper produced and heating does not help. The presence of lignin worsens the situation. Pulp from some bright coloured woods such as aspen, poplar and cotton wood serve the purpose of providing bulk. Chemical pretreatment of hardwoods yields pulp of strength properties equal to or better than mechanical pulps made from soft woods. Sulfonation helps lignin to swell and this provides additional fiber to fiber bonding strength.

Cold soda process is the oldest chemi-mechanical process. The wood chips are soaked in cold sodium hydroxide solution before refining. The temperature increases to about 80 degrees in half to two hours time. Hemicellulose swells in the alkaline ambient. If the temperature goes higher hemicellulose may dissolve. The energy and time requirement of refining is much less. The fibers are long and there are less numbers of fines. Increased chemical pretreatment will reduce opacity, yield and lower the brightness.

Sodium bisulfite treatment is the other pre-chemical treatment used on wood. The pH is retained in the 4 to 5 range. The sulfonation attacks lignin. If the pH exceeds 7 the pulp turns brownish. Cooking time is generally short. Sodium bisulfite in acid medium of pH 1.5 to 2 is impregnated into wood chips for about 60 minutes. This is followed by vapour phase digestion at 120°. This kind of acid sulphite pretreatment

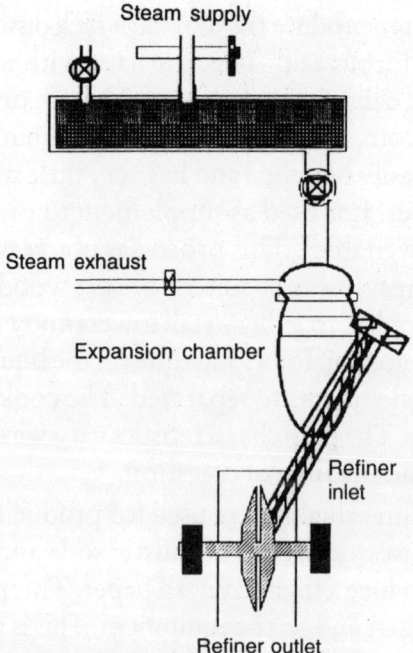

Steam supply

Steam exhaust

Expansion chamber

Refiner inlet

Refiner outlet

Fig. 2.5. Continuous pulping.

sulfonates the lignin about three times faster. Acid sulphite treated softwoods provide paper with better printing qualities, good opacity but low tear strength.

2.1.1.4 *Chemi-Thermo-Mechanical Pulping*

In Chemi-Thermo-Mechanical Pulping softening of chips is achieved by impregnating with a mixture of sodium bisulfite and sodium hydroxide. The pH is set between 9 and 10. The chips are treated in a steaming vessel to 130 to 170 deg. and then refined in a pressurized or atmospheric refiner. The properties of CTMP can be varied over a wide range by changing the parameters. The brightness of the pulp is increased with increase in sulphite.

Wood chips are cooked with alkaline or acid medium under

heat and pressure in chemical processes. In the sulphite process sulphur is burnt to produce the dioxide which dissolves in water to form the sulfurous acid. This is reacted with wood chips in the presence of calcium, sodium, magnesium or ammonium bisulfites. This compound solution dissolves lignin. Unbleached sulfite pulp is easily bleached and has very little strength when made into paper. It is used as supplement to the mechanical pulp to make writables. This process is not very suitable for high lignin chips. It is best suited for soft woods. The chips and liquid are cooked together in tall towers known as digesters. The treatment goes on for a long time till the liquid penetrates the chips and the fibers are separated. The cooking may last for thirty hours. The pulp characteristics vary with the strength of chemicals and the time of digestion.

For a long time alkalis were used for producing pulp. The kraft process uses a mixture of caustic soda and sulphate or sulphide to produce a high strength paper. This process is less severe on the fibers and on the equipment. This is more suitable for hard woods as well as other fiber sources such as esparto and straw. The waste chemicals are recovered. The success of the chemical processes is attributed to the successful recovery processes available. The process can also be automated and made continuous as in the Kamyr digester.

The Kamyr continuous digester has the following features: Wood chips are metered into the rather tall digester tanks filled with the appropriate chemicals. Certain amount of predigesting may happen and steam under pressure is applied to the chips as they enter. The cooking time may be very large in some cases. Pulp so produced comes in a virtually continuous stream. Screening and cleaning of pulp need be done to clean it up from shives and silica. Sedimentation or vortex cleaners may be used. Pulps may bleach hard or easy bleaching. Bleaching Kraft pulp is usually done by Peroxide bleaching. Hypochlorite and Chlorine Dioxide bleaching may also be employed.

Most of the pulp produced in US (about 2/3) is produced by the chemical process—the kraft or sulphide process. The yield of the process is about 40 to 45% of wood by weight. The pulp has very good strength properties and is used for very diverse applications. Unbleached pulp is dark brown (the colour of the grocery bags) and used in making bags, wrapping paper, container board and similar items. More than half the pulp made is bleached for use in high grade writing and printing papers. The sulphur released into the atmosphere produces obnoxious smell in the vicinity and is also a big polluter when released. Some of the chemical pulp may be added to newsprint to give additional strength.

A small amount of paper is made by the acid sulfite process, which produces a pulp that produces paper not as good as the kraft. The technique is slowly losing ground.

Semichemical pulp yields are higher giving 60 to 80% of wood as pulp. The process is mainly used for corrugating medium and board, newsprint and similar applications.

Ground wood pulp and the variations such as the TMP and CTMP now account for about 1/5 of US production. The yield is 90% of wood by weight. The built-in opacity and the large yield are big advantages. Tissue and toweling papers and newsprint are directly made from this kind of pulp. The mechanical strength is poor. Often the pulp is mixed with chemical pulp to add strength. The paper tends to degrade very fast in comparison to others.

(Deinked waste paper yield is about 80 to 90%. The paper may have to pass through many cleaning and filtration stages. The number of operations depends entirely on the source of the waste paper. At least in some cases the recycling means using paper from any of the woody sources directly. The pulp obtained is of very diverse quality and may be used in tissue products, paper board, newsprint and writing papers.)

2.1.2 Non-woody sources of pulp

2.1.2.1 Agricultural by-products such as bagasse

After crushing sugar cane and extracting the liquid part, bagasse is left behind. The main constituents of bagasse are:

- The rind including the epidermis, cortex and pericycle.
- The vascular fiber bundles comprising of thin-walled conducting cells with relatively thin-walled fibers with narrow lumen.
- Ground tissue or pith with fiber bundles distributed irregularly.

Crude bagasse contains 70 to 75% useful fiber and about 30 to 35% pith, dirt and other water soluble materials. Pith contains nonfibrous cells less than 0.4 mm in length. Removing the pith may be done in many ways. One of them is dry pithing which is breaking, blowing and screening to remove pith. This is quite inefficient. In moist depithing as the cane is being crushed in the sugar mill, the bundles are rubbed mechanically with disintegrating action to dislodge the embedded pith. The pith may be used as a fuel or cattle feed. When about 2/3 of the pith is removed, pulp of medium quality can be made. An advanced machine such as the one made by Swiss Puerto Rican Metallurgical Corporation is often used for depithing. This machine uses rotors to affect depithing. After depithing bagasse is ready for pulping.

Bagasse is pulped by any of the conventional techniques. Soda, sulphite and neutral sulphite are generally used. Pulp of bleachable grade is produced after digestion for 10 or 15 minutes at temperatures of 165 to 170°C. Digester pressure is maintained at 6.5 to 7 kg per cm^2. NaOH with 12% alkali, liquor to dry bagasse ratio of 1 : 3.5 is used.

(The Peacado Process, Cusi-san Cristobal process and many others have been developed specifically for pulping bagasse, as it is one of the important materials available as a by-product of another industrial process. The only limitation is bagasse is a seasonal product and hence storage is a serious problem.)

Bagasse pulps are easily bleached by the single hypochlorite process. Chlorine dioxide bleaching is necessary to bring brightness levels to the fullest.

Once depithed good quality pulp for newsprint applications can be made by mechanical processes, TMP or CMP. The Simon-Cusi process produces pulp of 62 brightness and a yield of the order of 45% on whole bagasse. Upto 95% of the pulp has been used in the fiber furnish to produce reasonable quality newsprint.

2.1.2.2 Cereal straw

Cereal straw can be an important raw material for pulp making. When manual collection is done it is possible to keep the straw clean by cutting a few inches above ground, but in automated units generally the entire plant is plucked out. There is generally deterioration in storage which has a bearing on the moisture content of straw. Straw has a low cellulose content but the holocellulose content is almost equal to that of wood, it has a higher pentosan content and little of lignin. Papers produced from wheat and rye straws are of good quality.

Pulp can be made by the mechano-chemical processes or the chemical processes. The soda process produces a coarse pulp good for boards, packing and wrapping papers. Bleaching may be done by three stage bleaching process.

2.1.2.3 Bamboo

The chemical constitution has starch, pectins, lignins and celluloses. Physically, it has the culm (the hollow portion) and the parenchyma (the ground tissue); the latter carrying fiber bundles. The quality of pulp obtained from some bamboos compares favourably with pulp from coniferous woods. The culm is joined by nodes. These nodes create a problem in pulping as they are denser and may contain siliceous impurities. Specially designed crushers are necessary to handle bamboo. Bamboo harvesting can be done between 4 and 6 years. The yield is about 3 tons per hectare per year.

Alkaline sulphate pulping is the best process for pulping bamboo. The acid process produces pulp of lower strength properties. The forest research institute in Dehradun (India) has produced a two stage digestion process which can be used on a commercial scale. Bleaching of bamboo pulp is difficult.

2.1.2.4 Reeds

The swampy delta areas in Egypt, Russia, Rumania, Iraq and some others grow certain varieties of reeds which grow to 3 to 5 meters in height and are about 2.5 cm diameter. These can be pulped easily on the sulphide base process.

2.1.2.5 Esparto grass

Esparto grass grows in southern Spain and Africa. Esparto pulp is suited for making fine quality printing paper. Esparto grass is composed largely of cutocellulose or pectocellulose. Digestion involves retention time of about 15 minutes at 165 degrees, with active alkali 15% and solid to liquor ratio of 1 to 3.5.

2.1.2.6 Hemp, Flax, Jute, Kenaf

Hemp, Flax and Jute are also agricultural products that can be pulped. Hemp pulp is particularly suited for making dense papers such as onion skin. Flax pulp is used for cigarette paper, condenser tissue, airmail paper and other expensive thin papers. Flax is generally digested in globe digesters with raw sulphur in soda. Indian linseed fibers when cooked with 10% alkali at 150 deg. for 4 hours yield 70% pulp. Jute pulp is used for making high strength bags, wrappings, drawing papers and tags.

Kenaf is an African plant. It is successfully grown in many

other parts of the world. It produces 3 annual yields. Kenaf reaches maturity in 100 days. The bast fibers yield long fiber pulp which can be blended with other fibers to produce high quality paper. The short fiber pulp from the woody core is used as a furnish for making heavy boards.

Most of the non-wood digestions are done in batch but continuous processes are also possible.

2.1.3 Secondary sources of fiber

2.1.3.1 Rags and cotton linters

Rags were one of the earliest raw materials for paper making. High grade cotton and linen rags are used for making the best grades of bond, writing and technical papers. No old rag can be used for the production of good quality paper. The major source is rag from textile mills and garment makers. The dyes that might be used must be strippable and the synthetics should be avoided. Rags must be cooked to remove, waxes, dirt, colourings etc. Usually rotary horizontal cookers are used. Sodium hydroxide, sodium carbonate and calcium hydroxide are the main chemicals used in rag pulp. Lime produces some water insoluble products and hence may be avoided. Washing and beating are simultaneous processes. The purpose of beating is to reduce the length of the fibers. The alpha-cellulose content is high. Rag paper is very strong and durable.

2.1.3.2 Waste paper and recyclable paper as a source

Waste paper is being used as a secondary source for many centuries although in a very small scale. In Japan even in 1000 AD paper was used as a source for pulp. The ink is bound to make it difficult to bleach the paper and invariably the paper produced is coloured. The fibers undergo a certain amount of breakdown because of the mechanical changes on the surface of paper.

The waste paper is classified into categories as:

1. Mixed waste paper—may contain molded matter, boxes, etc.
2. Corrugated waste—double lined kraft waste, used for production of liner boards.
3. Direct entry—white paper waste of generally similar brightness and quality has no prohibitive material.
4. Deinking grades—paper with ink in black or colour, the base can be used for making fine papers, book papers, envelopes and other kinds of tissue consumer products. Copying paper or paper with fused ink may have to be treated separately.
5. Newspapers not sunburned, standard inked.
6. Prohibitive or unusable material such as soiled or dirty paper.

Paper mills have always used their own waste—trimmings from cutters and rejected paper known as broke. The quality of pulp from the fibers which have undergone the process of paper formation once are not the same as fresh fibers but are an important source that should not be wasted.

Deinking is done by many techniques such as flotation, washing, centrifugation etc.

Washing: Washing the ink off means either separating the ink by dissolving it in a proper solvent or by colloidalising it with the help of soap-like emulsifiers. The inks are generally water insoluble once set on paper hence washing may have to be done with non-aqueous solvents. When emulsification is used as the separating process, the solubility is not important and the process is very similar to cleaning of cloth. The process is most effective for removing the sizing and some coatings on paper.

Flotation: Assumes that the solid ink is reasonably finely powdered and is insoluble in water. This applies to many of the fillers and additives. The process is very similar to froth flotation process used in ore extraction. The insoluble but heavy contaminant has an interfacial tension peculiar to its own and hence floats on froth. The froth having the contaminant ink is skimmed off from the surface or the liquid at the bottom is drained off.

Centrifugal Cleaning: Makes use of the density differences between pulp and solidified ink. The pulp is spun in specially built centrifuges. The ink and other impurities settle on the outer edge and are periodically removed.

2.1.3.3 Recycled fiber as source for paper making

The physical properties and mechanical properties of recycled pulp are very difficult to generalize. It depends almost entirely on the source and the amount and extent of deinking that is done. Deinking as mentioned earlier is understood as the process of removing all non-fibrous material contaminants from the waste. Hence the paper quality that can be produced is also quite unpredictable. The only solution is to make particularly large batches of waste pulp, mix it as necessary with some virgin pulp and make lab samples before producing actual paper. The amount of virgin pulp may be altered depending on the desired qualities of the paper.

Recycled fibers may yield papers with slightly higher bulk with a higher folding resistance against the grain direction. Generally recycled paper has lower folding problems and may have lesser tendency to curl. The reason and explanation for this may be that recycled pulp contains generally shorter fibers. The higher percentage of shorter fibers may occupy the space between longer fibers hence giving a better bulk and greater opacity. This is a particular advantage to the book publisher as lighter weight paper can be used without the problems of show through. The tendency of recycled paper not to curl and hence lay flat is also a good point.

One of the most difficult tasks of reusing paper is the difficulties of formation of paper. Usually the pulp tends to retain lumps. The tendency to have lumps is further aggravated because of inks that melt or diffuse and harden into paper. These lumps are difficult to break by the conventional methods. In fact the presence of lumps or opaque islands in paper is the tell tale sign of paper made from secondary sources.

The lumps show up on the surface of paper too—though not to the same extent as in bulk. The contribution of these lumps to the surface smoothness is significant but is not so important from the printing point of view.

The attitude of the printers to recycled paper is mixed. Most of them do not find any particular problem while some do feel that recycled paper is not as printable as virgin paper.

Linting and picking have been found to be more common in case of recycled paper. This may be because of a very large number of short fibers. This however is of importance only in the case of uncoated paper and is of no concern when paper is coated.

For uncoated recycled paper the ink mileage is expected to be marginally lower. Dot gain is also expected to be larger in uncoated recycled paper.

Brightness of recycled paper is marginally lower than virgin paper.

Permanence of recycled paper is very similar to standard paper.

2.1.4 Further treatment of fibers

The pulp if prepared at the site of logging is transported to the mill in bales, crumbs or slush pulp. The most convenient and economical way to transport pulp is chosen. When pulp is obtained at the mill it requires certain treatments before it can make stock or furnish suitable for paper making.

If the transportation is done as logs and if it is ground at the site the first of the following steps may be eliminated but the rest of them need be done.

The steps included are:

- Disintegration
- Beating and refining
- Stock blending
- Addition of nonfibrous additives
- Conversion from batch to continuous operation
- Stock metering and flow control
- Dilution to paper making consistency
- Screening and cleaning
- Defloculation.

The dry content of the fiber as received may be as high as 90% in air dry sheets. An equipment such as kneader-repulper or chest and rotor disintegrator is used for obtaining the pulp in a fluid transportable form in the desired consistency.

2.1.4.1 Beating and refining

Beating or refining is the next important step. Beating is a batch process while refining is a continuous process. The equipment used are hollanders. The results of both the operations are the same. The effect can be summarized as follows:

1. Length of fibers affected by the cutting action.
2. The surface of the fiber is split hence making sizing easier.
3. Internal fibrillation or bruising happens and the flexibility, swelling and plasticity are affected.
4. Extreme fibrillation may remove primary wall.
5. Curling and twisting of fibers may take place.
6. Redistribution of hydrogen bonds may happen due to fibrillation—not measurable.

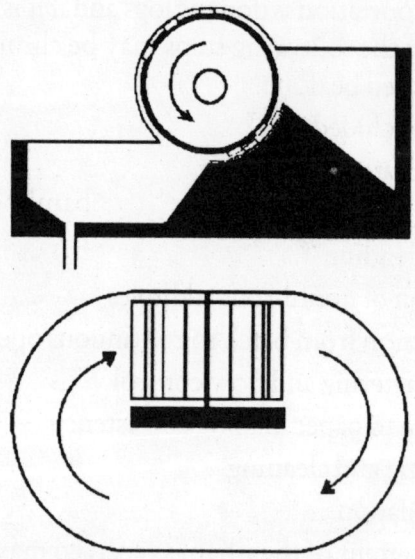

Fig. 2.6. Pulp beating machine.

7. Microcreping with fiber shortening is observed.

8. Fiber strength is not adversely affected.

The result of beating from the end product point of view are:

1. Strength and mechanical properties of paper.

2. Sheet formation properties, especially uniformity.

3. Density related properties.

Tensile, bursting and folding strengths reach a maximum with optimum beating and excessive beating either does not affect it or may deteriorate the mechanical properties.

Finding the optimum beating is difficult as the best process would be to make paper and check which takes time. Hence the only indication is to use the draining time of the pulp by squeezing a handful.

Higher the lignin content more difficult it is to beat the pulp. Glucose, starches etc. improve beatability. Sizing agents

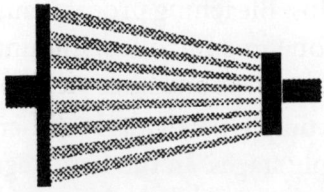

Fig. 2.7. Conical impeller of Hollander type refiner.

such as rosin and gelatin make it more difficult to beat. The beatability depends on pH and shows a minima or maxima. e.g. cotton pulp gives best strength when pulped at 8.5 while straw has a minimum between 7 and 7.5. Rate of beating is the higher at lower temperatures. Refiners work best at higher consistencies. However the process itself decides the consistency such as hollanders have 8 to 9% and refiners need 2 to 4%. However special refiners were built which used concentrated pulp at 40% or better to get good beating.

2.1.4.2 Bleaching of pulps

The colour of wood pulp ranges from cream to brown depending on its impurities. The purpose of bleaching is to get bright white colour. The bleaching process can be considered the process of removal of light absorbing substances. They may be lignin, lignin compounds, resins, fatty acid and esters, even hemicellulose in some cases. The impression of whiteness is still subjective and efforts to equate them to physical quantities have not been completely successful. The brightness often is indicated in terms of SCAN or ISO brightness numbers. This helps compare the relative efficiency of bleaching methods. The measurements are done on hand sheets produced and dried. The brightness of wood itself changes with storage due to some oxidization.

The earliest bleaching agent used was sunlight. Chemicals are used for making the lignin either soluble or to convert it to a colourless compound. Calcium hypochlorite is used for bleaching soda pulps. Bleaching process causes fiber damage and it is usually done in controlled amounts to maintain a balance between brightness and fiber yield.

Bleaching kraft pulp is particularly difficult. Kraft pulp is bleached in multiple stages. In the first stage chlorine gas is used which produces a partial degradation of lignin to make it water soluble. The pulp may be progressively bleached with calcium hypochlorite and then with chlorine dioxide. Computer controlled multistage bleaching produces a uniform quality pulp.

ISO brightness of over 80 is suitable for most applications.

In mechanical pulps most of the lignin is retained. Hydrogen peroxide, sodium peroxide and bisulfites are used as bleaching agents on mechanical pulps to keep the fiber yield high.

2.1.4.3 Sizing of paper

Sizing of paper is done in two ways. Engine sizing where the agent is added in the furnish or surface sizing when the chemical is applied on formed paper. Rosin is added to the pulp as a solution in dilute sodium hydroxide. When this is thoroughly mixed aluminum sulphate is added which precipitates aluminum rosinate on the active points of the fiber. Sizing prevents ink from spreading. Surface sizing is done with starch or carboxymethyl cellulose.

2.1.4.4 Additives and loadings

China clay, precipitated chalk, titanium dioxide and other fillers may be added to give opacity to paper. These fillers increase the smoothness, flatness, opacity and brightness of paper. They make paper more abrasive. Imitation art papers may contain 10% or more of fillers.

Addition of non-fibrous materials may happen which generally is addition of optical brightness and dyes. They are always added in solution form ensuring that they mix easily.

Blending is done of different kinds of fiber to get desired properties of the product.

The stock so made is to be moved to the headbox through pipelines. All that happens here is known as approach flow system.

The consistency is adjusted by dilution as desired to about 3 to 4%. Volume rate flow is adjusted.

Cleaning is done by the following four methods:

1. Separation by size, using screens.
2. Separation by settling rate—done usually by vortex cleaners.
3. Deaeration—removing of air bubbles—particularly the ones attached to the fibers (sometimes by evacuation).
4. Defloculation—open estranged fibers.

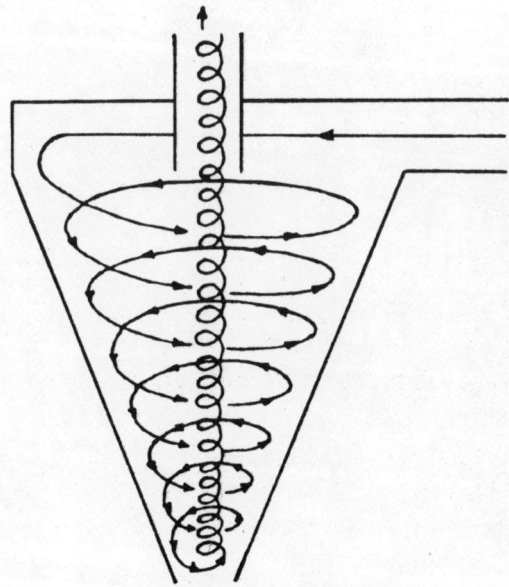

Fig. 2.8. Vortex cleaning of pulp.

After cleaning the pulp is subjected to further dilution. Dilution is set to about 1 percent. Only when the pulp is this thin, transporting pulp to the headbox and distributing it from the circular pipe to the shape of the slice in the headbox becomes possible.

Formation of Paper

3.1 THEORETICAL ASPECTS

The ease of formation of paper mainly depends on the amount of refining that takes place on the pulp. The beating process drives in more water into the fiber, opens up more of the fibrils and breaks them down. The resultant of a heavily beaten pulp is the grease-proof paper, which has higher transparency, settles much more densely with much less air between the fibers and is much stronger in its bond. On the other end is blotting paper which needs only a marginal amount of beating.

A sheet of paper is made from fibers in which some are coloured. When torn some fibers break and some others get extracted intact. Thus the strength of paper is decided by two independent factors, the fiber strength and the bonding strength. This is generally represented by a function such as

$$\frac{1}{T} = \frac{1}{F} + \frac{1}{B}$$

where T is the tensile strength index of paper; F is fiber strength index; and B is the bonding strength index.

Paper is a network of crossing fibers bonded to each other. The properties depend on:

1. The formation and density of paper.
2. Fiber orientation.
3. Relative bonded area.
4. Mechanical and strength properties of the fibers influenced mainly by the shape and source.

When bonding increases strength properties such as tensile strength and bursting strength increase. Folding endurance increases when fiber bonding increases but may fall if the bonding makes paper brittle. Opacity of paper decreases as fiber bonding increases. The two are related so closely that the optical measurements are considered a reliable way of estimating the bonding within paper. Calendering increases density with hardly any change in bonding strength. If the moisture content is low, the fiber bonding strength may fall due to calendering.

Interfiber bonding reduces surface fuzz.

Wet paper is weak and very dry paper is also weak. The strength peaks up and remains nearly flat over a large range of moisture content. The moisture would occur in paper in two distinct areas, the space between the fibers and within the fiber.

Paper leaves the fourdrinier wire at about 25 to 30% solids. At this time the paper can support its own weight over short lengths. The process of surface tension acting on the fibers starts at about 11 or 12% solids. The fibers are stuck together by water and the forces are minimal. The conditions of surface tension demand that the two surfaces are close together and have moisture but are not immersed in water. When the mat is pressed together and water is squeezed out, the forces between the fibers increases. If two flat surfaces 1 μm apart are held together by water the pressure is 145 kPa. Similarly for two round objects 2 μm apart the pressure is 3700 kPa. removing the interfiber water by sublimation leads to a highly porous, bulky sheet of high opacity and low strength.

Even small amount of surfactants (the ones that reduce surface tension) that may be included in the pulp by accident or deliberate lead to weak papers.

When the solids level is between 20 and 25% the thickness of the solid matters reduces very fast and the bonding strength increases drastically. This is because air intrusion takes place and the water content falls. After this range the fibers start loosing water from their body. The fibers swell in water and the amount of swelling—it is assumed—depends on the amount of beating that has taken place. It is observed that the fibers swell and in addition they undergo elongation. The elongation may be as high as 25% in some types of fibers. All this is due to water retained in the fiber. Loss of water adds to the strength of the fiber and hence that of the sheet. Often these last traces of water need be extracted by external pressures rather than internal draining.

The mechanism and reason for bonding strength between fibers has been discussed in detail in literature. Paper does not have any ionic bonds. Paper is not as strong as plastic of similar dimensions. Plastic has primary valence bonds. The strength of these bonds is in the region of 30 kcal/mole. Hydrogen bonds have strength in the order of 5 kcal/mole and other polar bonds about 2 kcal/mole. The bonds in paper tend to look more like hydrogen bonds. $-OH$ and $-NH_2$ groups are the ones which form hydrogen bonds. $-OH$ group is in plenty in paper and it can easily come to distance as low as 0.25 nm i.e. atomic distances to form hydrogen bonds.

The partial solubility theory approaches the problem in a different way. Cellulose is supposed to form crystallites on the surface by dissolving in water under certain conditions. It can also be assumed as a two dimensional colloidal system which is anchored to the fiber in the third dimension.

Commercial pulps have varying lignin content. It is about 1.5% for soft pulps, 8% for hard pulps and about 16% for semi-chemical pulps. The effect of the lignin concentration on the strength of many pulps has been studied on hand sheets. The burst and tear strength fall almost proportionately from 2% to 20%. The fold strength decreases very rapidly from 2% to 15% and then flattens out at a low value.

Recycled paper, especially the clean ones or the mill brokes pass through the drying and moistening cycle a second time. The quality of the pulp does not remain the same and the properties are not reversible. An increase in stiffness and decrease in flexibility is observed.

Addition of many a substance to the fibrous furnish reduces the fiber bonding strength. Hydrophobic substances such as rosin and paraffin certainly affect the strength of paper. Hydrophilic substances such as clay and alumina lead to loss of strength to a lesser extent.

A few substances such as high molecular weight hydrophobic colloids such as starch, proteins, vegetable gums, water soluble resins (Polymers such as acrylamide) increase the bonding strength. These materials might act as fiber to fiber adhesives.

The pH has little influence on the strength of the paper. The acidity decides the permanence of paper.

Synthetic resin fibers, glass fibers and asbestos are added sometimes in paper. Synthetic fibers do not disperse themselves well in aqueous media and do not bond to form a sheet of paper. Bonding agents are generally used in such cases.

Water is essential for the plasticization of cellulose. Cellulose fibers do not spontaneously dissolve in water. Beating and fibrillation is the only way that has a direct relation to the strength of the finished paper. Poly functional molecules add to wet strength of paper. Low molecular weight polymers such as urea and formaldehyde add to the wet strength.

3.2 SHEET FORMATION

Sheet formation refers to (1) the mechanism of formation of the sheet; and (2) the qualitative appearance of the formed sheet. The second depends on the fiber orientation and the degree of packing of solid fraction. The strength of paper— both tensile and tearing strength, the amount of expansion, contraction happening at different moisture levels also depends on formation.

In early days of paper making the machines operated at low speeds. These days, excepting for the hand made variety, the machines work at very high speeds and hence the entire mechanism of formation is much more under strain. The mechanism of paper formation consists of the fourdrinier where a certain section of a wire mesh receives the wet pulp, the water is drained initially and subsequently pressed and formed, dried and calendered or super calendered.

The steps of paper formation are:

- The headbox
- The forming section
- Sheet transfer from forming to pressing
- The press section, the dryer section and the finishing section are the subgroups of paper formation.

The pulp at the headbox has consistency of the order of 0.2 to 1%. Before pressing at the wet end it is 18 to 23%. At the end of the press section it is 33 to 50% and after drying water is brought to 92 to 96%.

Most of the problems of a paper plant is in the use of very large volumes of water to suspend the pulp and dry it. This is to reduce the water level to a tolerable limit on paper. In a typical newsprint plant manufacturing 2500 kg of newsprint per sec, the initial content of about 90,000 liters per second is brought down to 2500 liters per second of water.

Flocculation: Fiber sometimes tends to form flocs which

must be avoided. This is generally due to the shape and size of the fibers themselves but also affected by hydrodynamic variables of the medium such as turbulence, rate of shear, time etc. Dispersing action is affected by additives. Eddy currents of low velocity, low consistency and large fiber length are the main causes of flocculation.

Additives and loadings have their own particle sizes e.g. TiO_2 0.2-0.5, $CaCO_3$ 0.2-05, Clay 0.5-1.0, Talc 1.0-10.0 (all sizes in μm) It is important that these are held within the fiber mat for any effective results.

3.3 DRAINAGE OF FIBERS ON PAPER MACHINE

Pulp moves continuously through the slip onto the moving belt. The water level at this instance is very high. Water drains through the mat which keeps building up progressively. In a simple case the rate of volumetric flow Q through a mat of filter medium can be represented as

$$Q = \frac{K\Delta PA}{\mu L}$$

where K is a constant of proportionality which can be named specific permeability, ΔP is the pressure differential across the bed in the direction of flow, A is the cross sectional area and μ is the viscosity of the fluid and L the bed depth.

This equation has undergone many changes to make it more practicable though by itself it gives a complete picture of a static medium. The real systems are many orders more dynamic hence only approximations are done, one of them the following way:

$$\frac{Q}{A} = V_s = \frac{1}{\mu R}\frac{\Delta P}{w}$$

where V_s the measured quantity is the flow rate per unit area and w is the basis weight of the mat and R is the specific filtration resistance of the mat. R is dependent closely on the surface area of fibers to a specific mass and volume per unit mass of fibers.

Retention: The fiber is flexible when wet. When passing over the wire or mesh quite few pieces of the fiber are retained on the wire and the smaller fibers directly pass through the formed mat. Hence depending on the forces the wire side is going to be less rich in longer fibers. This can be considered to be related to the ratio of grid spacing and fiber length. It depends on the geometry of the spacing of wires.

Dynamic forces in the semi-formed mat: During the filtration process the fiber mat deposited is subjected to drag forces by the liquid flow. The flow rate through the fiber mat hence is also calculated on the basis of compressive studies on the mat.

$$C = C_0 + MPe^N$$

where C is the fiber concentration and M, N are empirical constants.

M depends on the kind and type of fiber used and the amount of additives and impurities that may exist in the stock. The value of M varies between 0.001 and 0.015, and N between 0.15 and 0.4.

3.4 SURFACE TENSION EFFECTS

In the thickening process the surface tension effects on fiber are also important. The numerical relation is as ΔP the pressure difference between air-water interface is given by

$$\Delta P = \frac{4\sigma \cos \theta}{r}$$

σ is the surface tension and θ is the contact angle. r is the radius within which an equilibrium condition is reached.

3.5 PRESSING THE WET WEB

When the wet web leaves the couch roller and enters the press section, it comes into contact with a felt pickup mechanism that moves the web to the nip of the press. The pressures on the mat saturate the web with water and the water escapes onto the felt. When under the roller nip water moves out at the highest flow rate in the thickness of the mat. Once it leaves the nip the two sides expand and water diffuse out from the center. The felt is more resilient hence attains its original shape very fast, whereas the paper mat takes more time to do the same. Upon separation from the felt the mat redistributes the water and usually a moisture level greater than what was faced at the center of the nip results. This effect and the efficiency of pressing out the water depends mainly on the quality of the fiber and the porosity of the mat which in turn depends on the amount of refining treatment the fiber had.

Generally wet pressing increases the fiber bonding, and increases the strength of paper. Highly refined fibers hardly get affected.

SGW papers are less compressible in comparison to sulphate pulps. This is because the fiber is generally shorter and is already bunched. However by heavy refining it is possible to change the characteristics.

3.6 CALENDERING

After the sheet is dried it is passed through calendering rollers. These are a stack of heavy metal rollers which apply a lot of compression pressure on paper. The resultant effects of this on paper are as follows:

- The pressure smoothens the felt and wire marks.
- Binds any loose fibers back to paper.
- Remove any unevenness and cockle in paper.
- Level off any lump formation.

Calendering increases the density by reducing the thickness of paper. The opacity of the paper is reduced. The sheet expand in both directions. The tensile strength is marginally increased.

3.7 THE DRYING PROCESS

Paper leaves the press section with about 35 percent solids. This must change to water percentage in the range of 92 to 98%.

Paper is dried by many processes. Hand-made paper is dried by loft drying. Webs may be dried by festoon drying or on barber dryers. Most machine made paper is dried by passing the wet paper over steam heated rollers.

In practice the drying is not a steady state process. The paper heats when it comes in contact with the first roller and cools down when it leaves. Throughout the drying process paper is subjected to alternate cooling and heating cycles. When wet paper touches a heated roller the water at the interface evaporates. The pressure at the interface increases and forces the water to flow back on the web. As evaporation continues water is also flushed from the interior to the surface. The condition can be represented by modified Ficks law.

Water is held within the sheet in several forms. 1 to 2 percent of the water is held by the cellulose molecules mono-molecularly. It is extremely difficult to take this water out. About a quarter of the water is held as imbibed water by the capillary forces existing between the cellulose fibers. Largest amount of water is held between the pores as free water. This is the easiest to remove.

During the early stages of drying the fibers are free to slide over one another. As the water is driven off, the fibers are drawn closer together and stronger bonding take place. Once the critical drying point is reached the shrinkage begins to take place and bonding begins.

Once the solids content exceeds about 80%, there is no appreciable change in strength due to reduction of water.

Shrinkage takes place during drying in the cross direction. The shrinkage is the greatest on the edges. Excessive cross directional shrinkage causes a series of raised ribs in the machine direction. A condition known as grainy edges also occurs.

When Yankee dryer is used for producing machine glazed papers, a highly polished surface and finish is got by sacrificing the strength of the paper.

The paper web continuously undergoes forces of elongation of the web. Even within the dryer there is the lengthwise tension due to draw. The effect of such linear increase results in causing the following effects on paper:

- Reduction of basis weight.
- Increase in compactness.
- Reduction in burst strength.
- Increase in tensile strength.
- Reduction of stretch.

(Conversely hand-made papers made under absolutely unrestrained conditions and the free angular distribution of fibers make them what they are.)

In manufacture, the paper is known to suffer the following levels of elongation:

Couch and first press	2%
1st and 2nd press	1%
3rd press and dryers	0.5%
Dryers and calenders	1%
Calenders and reel	0.5%
Total	6%

If paper dries under tension, rigidity of the paper increases leading to a lowering of bursting strength. The same stock dried under tension has about 15% lower bursting strength compared to the one dried in free conditions. Similarly the physical properties of the paper from edges of a web are bound to be different than the one in the center because the web certainly is free as one moves towards the sides. Paper should be subjected to slow rise in temperatures at the wet end of the dryer so that picking, blistering, blowing, case hardening, cockling and curling are reduced.

Each paper machine has an optimum speed at which a particular kind of stock would perform best. In general most of the machines work at the highest speed for tissues and slows down for newsprint and kraft sheets.

CHAPTER

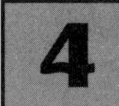

The
Paper Machine

4.1 THE PAPER MACHINE

The paper machine converts the prepared liquid suspension of fibrous material—pulp—into solid-paper.

The pulp is sent to the paper machine through pipes as if it is a liquid. The furnish is passed through filter screens and then delivered on to the head box with a centrifugal pump. The thin opening in the head box through which the material has to flow is known as the slice. The stock should deliver to the slice uniformly. The uniformity should be confirmed in respect of distribution of fibers, consistency, velocity and volume. These are the factors that decide the quality of paper produced and also the speed with which it can be made. As machine speeds and widths have increased, the head box has essentially become a series of hydraulic nozzles that distribute pulp uniformly over the width.

A reasonable turbulence must be available at the slice. This is obtained and controlled by designing the internal geometry of the head box. Very many designs have been experimented with. In early days simple gravity head feed was used. The present trend is to have pressurized feed instead of gravity feed at the head box. Generally the stock contains about 99.5% water and half percent solids at the head box.

The wire on to which the stock is delivered is the main drainage element of the paper machine. It is there throughout the wet end and has the greatest influence on the formation of paper. The wire is continuous and endless. In the early days of paper making, phosphor bronze wires were woven to make the wire net belt. Plastics are used now a days and are longer lasting and better than their metallic counterparts. Synthetic wires form a more even sheet of paper.

The width of the wire is known as the deckle. The deckle widths vary from a few meters to about eight meters. There are two kinds of fourdriniers. The single wire and the double wire type. The double wire was initially made for board making, to permit carrying more furnish. This principle of having a second overhead wire in the so called duoformers or inverformers has a few advantages over conventional fourdriniers and hence is used in making fine papers as well as boards.

On fourdrinier like open single wire formers the forming time is large and the turbulence created by the table rollers overshadows all other forces. Consistency becomes more important at higher speeds. In twin wire formers with suction breast rolls the formation is instantaneous with very little or no disturbance after the head box. In such cases the formation and other physical properties are decided by the fiber dispersion in the jet. The twin wire formers demand the most from the headbox designer. The most important factor that affects the uniform distribution of stock is air in the stock. Excessive air produces large variations in basis weight (weight of a ream of 500 sheets of a standard size of paper.) and may lead to mottling and surging. Air may form foams in the stock. Foam is nothing but tiny air bubbles size of 1 μm held together. If it is strong it may cause many problems in paper. Deaeration may be necessary in some cases of stock to remove all possible air bubbles.

It is generally believed that the fibers tend to align themselves with the direction of motion of the fluid in which it is suspended. This is only partially true. The movement of a fiber in a mobile stream may be random. However alignment does take place if there is acceleration in the flow.

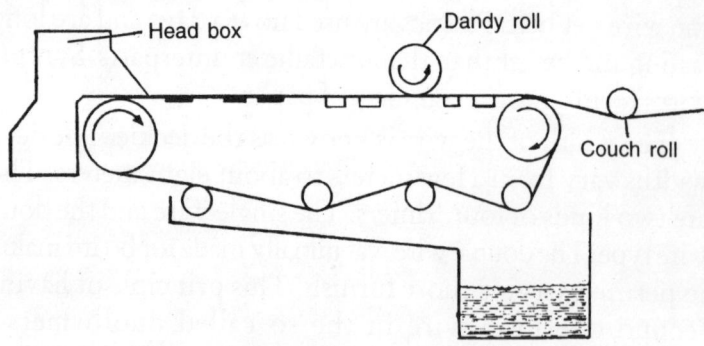

Fig. 4.1. Wet end of fourdrinier machine.

The figure shows the wet end and the wire portion of a conventional fourdrinier. The most important thing to note is that the furnish that enters the headbox at concentrations as low as half percent leaves the wire at about 20% solids. This means that about 98% of the water is drained off in this very stage. As mentioned the chances that the short and thin fibers getting lost is maximum at this part of the operation.

The head required for feeding the pulp through the slice was provided only by gravity, by maintaining the level of the pulp in the head box. The furnish was delivered onto the specially made wire through the slice. The endless belt of wire was made of phosphor bronze in the early days but is of plastics in all modern machines. The pattern of weaving is also important depending on the kind of paper that is to be made. In all modern machines the headbox has built in pressure to

- Reduce flocculation

- Reduce foaming. The stuff is kept circulated usually by the flow control itself.

The discharge of stock Q at the slice can be represented as

$$Q = blv$$

where b = breadth of the slice, l = the length of the slice, and v = the velocity of stock in meters per second. This obviously belongs to the head that can be provided in the box and also g the acceleration due to gravity if external pressure is not used. The angle of the slice with the wire is also of importance as this affects the flow of stock in many ways.

The relationship between the velocity of flow and the velocity of wire decides how the successive layers are built in the sheet. Some paper makers think that this is the single most important factor in paper formation. In both single wire and double wire forming machines, the strength properties both in the machine and cross machine direction are affected by this relative speed. The properties of sheet release from wire and retention are also affected.

If the wire speed is in excess of jet speed backward waves are formed and if it is otherwise forward waves are formed in the sheet. A combing effect may also be observed. Though these changes are observable only at large differences, it is best to maintain both the velocities at nearly the same value. If the wire speed is much higher it is likely that more of the fibers will align themselves in the machine direction. The difference depends on many factors such as head box design, fiber length, consistency etc.

The early paper machines applied a shake side ways. The effect of the shake is reduced at higher machine speeds. At about 1000 ft. a sec the effect is nearly zero. This action known as close up helps in breaking up the fiber flocs. Effectiveness of shake is related to the velocity difference between wire and flow and the shake speed.

The underside of the wire has foils and baffles arranged in specific ways. They direct the flow of draining water. When arranged correctly they produce a suction and a vacuum is created by the Bernoulli principle. The baffles can be so designed and placed in such a way that faster speeds of the wire produce a greater amount of drainage.

When the water level is in the order of 97.5% that is the paper is hardly formed the dandy roll makes the impression on the film of pulp deposited. The dandy roll is a hollow wire mesh roll generally of texture very similar to the paper making wire. The dandy rolls have patterns engraved on the outer surface. These leave water marks on paper still under formation. It is also possible to set certain patterns on the paper with the dandy rolls. The wove structure or the laid structure can be introduced on the damp stock. The paper mark is not a deformity of the paper in that the grammage remains unaffected. The fibers get reoriented in due to the water mark on the semiformed paper.

After the dandy roll almost 80% of the water that was in the original is removed. The formed paper moves onto the suction boxes where more of water is removed and the level solid increases to about 17%. The mat is compacted. The number of suction boxes may vary from 2 to 10 and the vacuum level is 7 to 37 kPa. Lower vacuum is applied at the first box. Too much of suction is also not considered good. The lump breaker and the suction couch (operated at 45 to 75 kPa vacuum levels) do further extraction of water and the paper at about 22% solids can jump small distances on its own.

In the conventional fourdrinier the couch roll was the only driven roll in the wire section. It pulled the wire which in turn pulled all the other rollers.

Twin wire formers that work on the principle of injecting a jet at the nip of two moving wires, allows for fast drainage and well dispersed fiber suspension. They provide symmetrical sheet properties on either side of paper. There are many kinds of twin wire forming machines available. Each of them is specially made for certain kinds of application and is meant for producing a certain thickness of paper from a certain variety of pulp.

Conventional fourdriniers at the best are used directly as tissue forming machines. However most of the quality paper is made on the duo formers. A substantial amount of paper is made from machines that operate as board making machines.

The couch if assisted by vacuum removes a sizable amount of water from the mat before it is transferred to the drying section. The vacuum assisted couch compared to the pressure couch gives a larger working life to the wire as well. The transfer may be assisted with a small blast of compressed air or in some cases the drying unit may be placed very close to the wire section. This distance to which the mat can stand by itself is a property of the pulp. Often in cases such as ground wood pulp (e.g. for newsprint) this distance may be very small. The distance to jump is about 2 cm.

In drying the wet mat is pressed between two felt belts. The operation only applies pressure and simultaneously introduces a certain amount of compacting. The process itself has been studied quite deeply. At a point in pressing all parts of the system have an equal amount of moisture. When the squeeze is released there is a redistribution of water within the system. The purpose is to remove as much of water as possible by pressure as this is much cheaper than drying paper by heat. The number of presses are 3 to 4. They may be simple metal on top and rubber at bottom kind of machines or may be again vacuum assisted to make water removal more effective. The last press usually is a reversing press where the web passes in a direction opposite to the nip. This helps improve over the two sidedness of paper.

Many kinds and types of presses are used and the efficiency

of drying depends besides other things on the quality of felt and the arrangement for heating the rolls. There are five different kinds used. Plain press, suction press, grooved-roll press, fabric press and high intensity press. felt parameters such as saturated moisture, vacuum dewatering, air permeability etc. are important.

4.2 THE DRYING SECTION

Paper from the press moves to the drying section. This is the process of removing the last traces of unwanted water by evaporation.

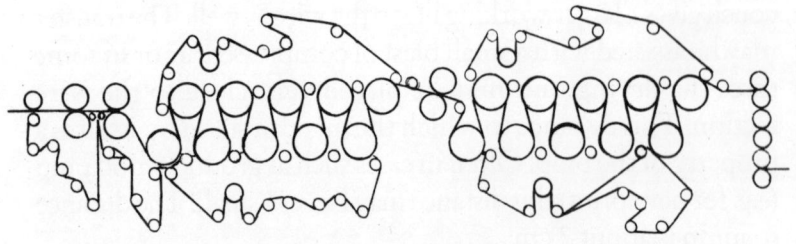

Fig. 4.2. Drying and calendering section.

In the simplest of the processes i.e. the Yankee dryer or the Machine Glazer roller the wet paper is pressed and dried at the same time. After minimum of pressing the wet web is pressed on to a highly polished very large diameter drum roller. This roller is heated with steam to a high temperature. A set of flexible rubber rollers press the paper very hard onto the drum. A cast of the surface is produced on paper. The paper is scraped off with doctor blades from the glazing drum. The paper is dried and glazed at the same time. This finish is often seen on one side of kraft paper as the Yankee finish.

In the normal dryers the paper from the press is moved over rollers which are covered with two belts of felt. The paper is dried slowly by increasing the temperature of the steam heated rollers gradually. The paper web may pass between two heavy

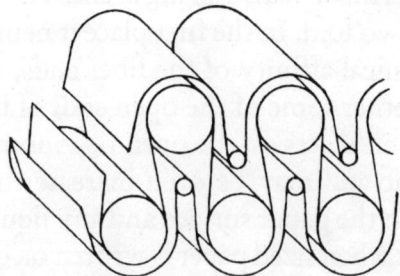

Fig. 4.3. Felt dryers.

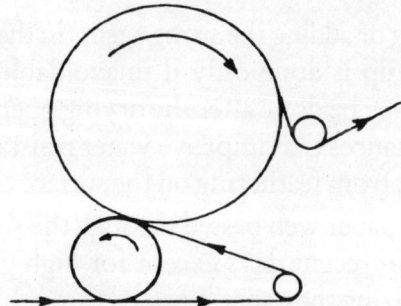

Fig. 4.4. Glazing of paper.

glazing rollers in the intermediate stages of drying, which improves the surface finish of paper. Calendering may also be done as a step in-between.

Drying may be done by many more techniques depending on the kind of stock and other technical requirements. Cylinder drying—transfer of heat from heated cylinders to paper, Yankee cylinder drying, impingement drying—having a hood containing high velocity high temperature air to assist drying, through air drying—where highly porous media may be dried with hot air, infrared drying etc. may be used.

4.3 SIZING

Sizing is a chemical treatment that renders some desirable properties to paper. The paper is either sprayed with or dipped in or some times made with a sizing agent. The action of the sizing agent is two fold. In the first place it neutralizes to an extent the chemical affinity of the fiber ends, when it may even bring together some of the open ends of fiber strands. The sizing chemical acts on the open regions making paper less hygroscopic and may be even increase the interfacial tension between the paper surface and any liquid on it. The net effect is that when sized paper is written on or printed on much less of spread is observed. This is most desirable when very thin inks such as used in writing or ink jet printing are used.

Engine sizing or adding the sizing agents in the preparatory stages of the pulp is done only if unavoidable. Otherwise surface sizing may be done after the drying operation. Sizing agents are substances that improve water resistance of paper and prevent ink from feathering on the surface of paper.

Tub sizing—paper web passed through the sizing agents— is rarely done in recent days except for high grade writing papers. In tub sizing the paper is passed through a solution of 5% to 7% gelatin with alum as hardener. Formalin may be added as a preservative. The excess of the size is removed by

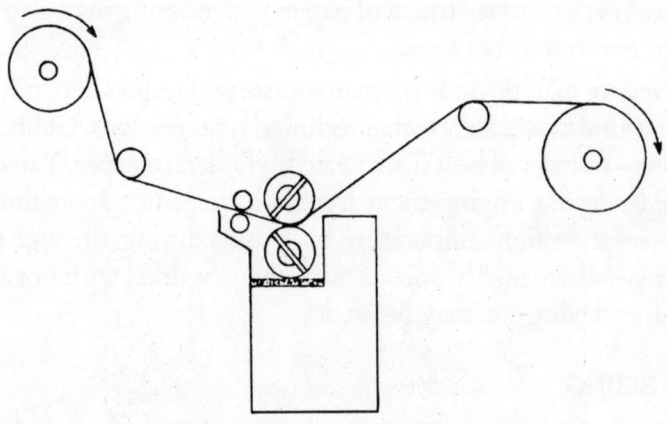

Fig. 4.5. Tub sizing of paper.

rubber squeezing rolls and then slowly dried. Mostly it is surface sizing—applying the sizing agent through roller coaters—is done for almost all kinds of papers. In the early days this was an off-line process. Recent trend is to have the entire operation in one go over the web. There are two variations i.e. the vertical size press and the horizontal size press. Besides other things the surface sizing improves the bonding properties and hence reduces drastically the loose fibers—fluthing.

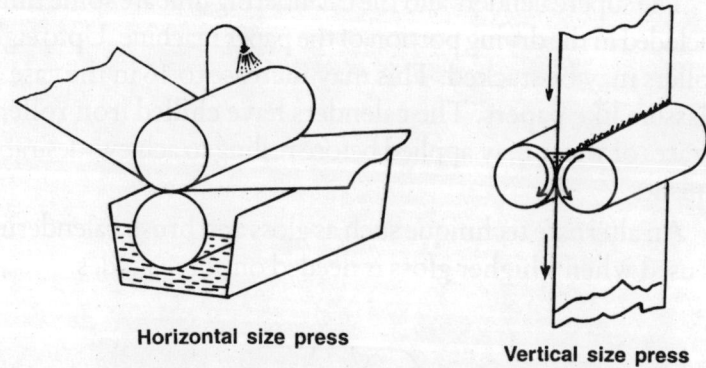

Horizontal size press

Vertical size press

Fig. 4.6. Horizontal and vertical sizing units.

Size may be sprayed on both sides of the paper web by a series of nozzles or jets and then leveled. China clay mix may be applied almost to 4 gms/meter squared per side which adds opacity and printability.

The most commonly used sizing solution is starch. Other materials such as gelatin, carboxymethylcellulose (CMC), polyvinyl alcohol (PVA) and wax emulsions may be used.

4.4 CALENDERING

After the paper or board is made in its basic form, a lot of other finishing operations may be performed. The paper web may be continuously monitored for defects and omissions. The finishing operations can be grouped as calendering and coating operations.

Pressing paper or board between large heavy polished rollers affects the surface of paper. Smoothness and gloss are increased. If a moderate finish is acceptable it may be done directly on the paper machine but if a high finish is required it may have to be done by super calendering in a separate step. While calendering depends only on pressure, super calendering depends on pressure and friction to achieve the desired level of finish.

The supercalenders and the calendering unit are some times included in the drying portion of the paper machine. Upto eight rollers may be stacked. This may increase to 16 in the case of glassine like papers. The calenders have chilled iron rollers. Traces of water may applied before rolling to achieve desirable finishes.

An alternate technique such as gloss and brush calendering is used when a higher gloss is needed on some grades.

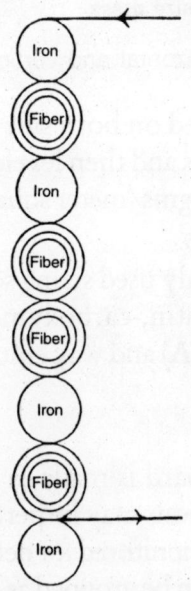

Fig. 4.7. Supercalender rollers.

The basic calender can give paper a fairly smooth finish known as Machine Finish. This varies depending on the stock used and the degree of pressure applied. A full MF can be considered semi-smooth and gives satisfactory reproduction of half tones.

Supercalenders work at nip pressures in the neighborhood of 250 kN/m^2. The kind of polish depends on the speed of the paper and the pressure. A considerable amount of skill is required in setting the variables to get desired results. The finish is very smooth and has high gloss. This is known as SC finish. The supercalender stack is separate from the fourdrinier so that the web can be fed with greater precision. These machines generally run at speeds higher than fourdrinier.

The entire operation of calendering is mechanical. It damages both the surface and the interior of the paper material. The effect is more pronounced on the surface. Fibers are broken and pressed on to the surface making the surface less porous. There is a certain amount of leveling that takes place. The valleys on the paper surface are filled with broken pieces of the fiber. Lot of dust may be created especially when the paper is not treated any further.

The bulk of the paper is also affected. The basic operation is a non-reversible compression of the bulk. This may mean certain amount of physical damages to the fibers or realignments of fiber bundles. The density of calendered paper hence is greater than before calendering.

The overall absorption of liquids on the surface is changed greatly because of the surface and the body changes. The paper turns less porous and hence the space available for liquids is much less both on the surface and the body of paper. Ink gives a much larger mileage and has less of a spread on calendered paper.

Coated papers are often super calendered to give gloss art paper.

Supercalendering is an expensive process and hence is not quite a preferred process.

4.5 COATING

The printing properties of paper can be greatly improved by applying a thin surface coat of non cellulosic material. This may a mineral pigment and a binder. The surface of paper as it is composed from many fibers is neither uniform nor fully even. Hence this forms a perfect base for a layer of coating. The finely divided pigment fills the pores and increases tremendously the surface finish, colour, gloss, absorptivity etc. Most of the imperfections are covered by the coating. The paper base is used simply as a carrier.

The coating mixture consists of two parts—the vehicle or binder and the pigment. The pigment is very similar to the loadings that could be added to the pulp. These inorganic materials usually are naturally occurring china clay, calcium carbonate, calcium sulphate, barium sulphate, titanium dioxide, alumina etc. The pigment may be used as it is or in the form of mixtures. They are ground well and are dispersed in the binding material. The adhesives or binders used may be natural materials such as starch, casein etc. or synthetic materials such as styrene butadiene, butadiene acrylonitrile etc. The coating mixture is ground to a consistency of milk and may have about seven

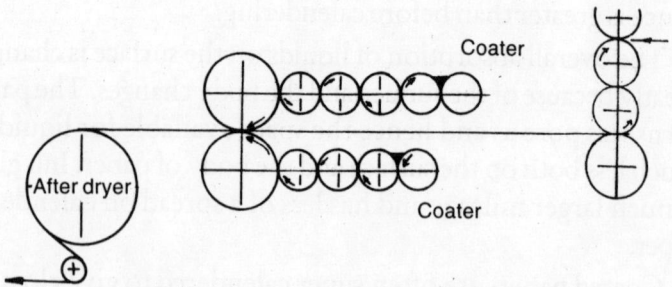

Fig. 4.8. On machine roll coater.

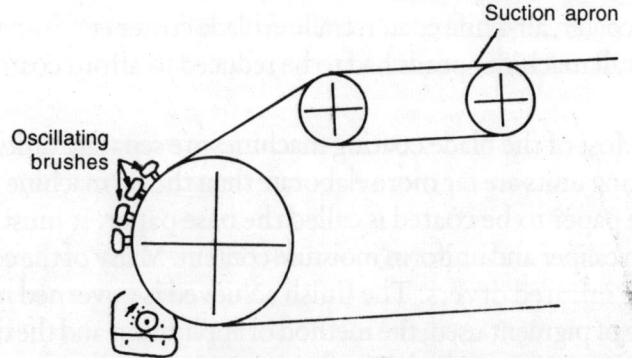

Fig. 4.9. System of brush coating of paper.

parts of vehicle to pigment. Other additives such as antifoaming agents, dyes and hardener such as formaldehyde may be added.

Although a coating can be applied on any surface the paper base must have reasonably good mechanical properties before coating can be done.

Many techniques have been developed to coat paper. The main processes are on machine coating and off machine coating.

It is possible to incorporate a coating unit after the drying operation right on the fourdrinier. These include the champion

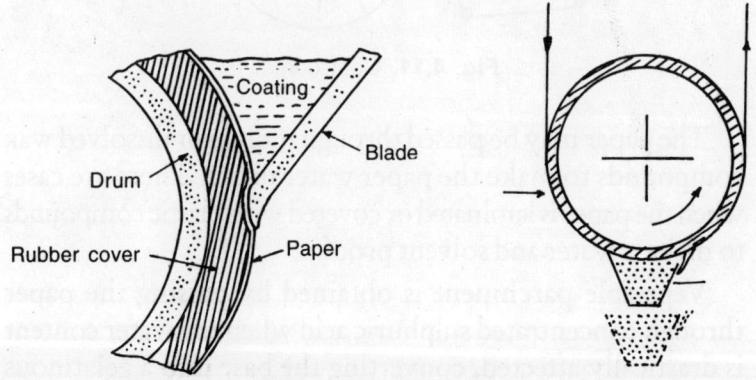

Fig. 4.10. Blade and flexible coating systems.

bar coater, air-knife coater, trailing blade coater etc. Sometimes overall machine speeds had to be reduced to afford coating on line.

Most of the blade coating machines are separate units. The coating units are far more elaborate than the on machine units. The paper to be coated is called the base paper. It must be of even caliper and uniform moisture content. Many of the coaters have infrared dryers. The finish achieved is governed by the type of pigment used, the method of application and the degree of calendering applied. The finish may be matt, semi-smooth or high gloss. Embossing may be done at this stage on the paper. In the figure we can see the blade coating and the inert blade coating which are two of the most popular techniques.

If an ultra-smooth surface is required the coated paper may be supercalendered or glazed.

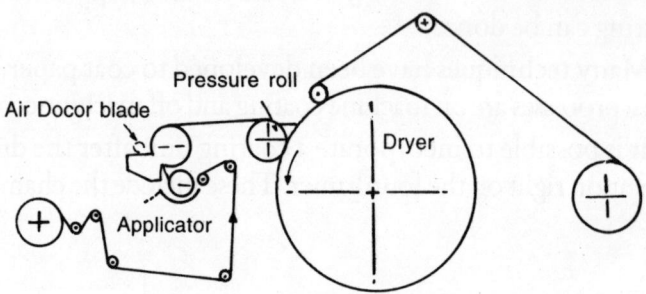

Fig. 4.11. Cast coater.

The paper may be passed through molten or dissolved wax compounds to make the paper water proof. There are cases when the paper is laminated or covered with plastic compounds to make it water and solvent proof.

Vegetable parchment is obtained by passing the paper through concentrated sulphuric acid when the water content is drastically affected, converting the base into a gelatinous state.

4.6 FINISHING OPERATIONS ON PAPER ROLLS

The width of the paper reels as they come out from the paper machine are anywhere between 1.5 and 8.5 meters. Generally this is wider than the calendering machines and casting machines. The reel is known as the jumbo reel weighing in excess of 20 tonnes. The jumbo reel is slit and rewound into smaller coils. To meet the demands of fast running printing presses, running at 5000 meters per minute paper must be wound with even tension. This can not be achieved on the jumbo reel and hence rewinding to smaller reels of workable diameter and width is done. Any joins due to breakage are rejected. Footage will have to be well made to stand the tension and speed of the printing operation.

Slitting is the first process in which these reels are sliced into smaller ones. It is combined with trimming—removing of the two rough deckle edges of the reel. A pair of adjustable circular knives perform the operations of slitting and trimming over a cutting table from where paper is rewound on cardboard centers. Many reels of the smaller size are wound together. The shavings of the deckle edge are repulped. The whole operation is done at a very high speed and if no further finishing necessary then the reels are wound under controlled tension conditions.

Cutting the reels to sheets is a separate operation. If again no finishing operation is necessary then it can be combined with trimming and slitting and no rereeling is necessary.

In the cutting operation the web passes over a fixed knife running across the width of paper. The cut is made by the chop knife which shears the paper.

Mostly the cutters are fed with slit reels. Electronic sensing is done to get good quality sheets. The sheets so cut are stacked with automatic counters and inserts to indicate 100 or 500 sheet marks.

The papers may be conditioned or brought to the same moisture condition as it might get after the first exposure and print to ensure that the dimensional stability is maintained especially in colour print.

Board and Hand Made Paper

5.1 BOARD MAKING

Board technically is a variation of paper with the following characteristics and differences.

1. Board is thicker—has a higher grammage. This is in excess of 120 gsm.

2. Board generally is opaque. There is no special process necessary to ensure opacity.

3. Board is generally stiffer.

4. Inferior quality pulp may be used in making boards as printability or ability to write is not very important. However boards may be used uncoated in some packaging applications.

5. Techniques are available for making multiply boards i.e. boards with different interiors than the outer surfaces.

6. Boards are generally made on smaller deckle and slower machines. The large amount of pulp that has to be delivered to a small area makes the invention of new technologies very different.

Conventional fourdrinier is used mostly direct as a tissue forming machine. However most of the quality paper is made on the duo formers. A substantial amount of paper is made from machines that operate as board making machines. The inverform machine may be used as it is for making boards.

The pulp making for board generally follows the same procedures as paper. The volume of water that has to be drained is much larger. More fibers have to be deposited on the mat than in the case of paper. Hence the pulp is only lightly beaten or refined. If pulp from waste paper is used almost no refining is needed, only a thorough cleaning and disintegration is ensured.

The cylinder machine for board manufacture was patented by John Dickinson in 1809. A wire screen cylinder revolves in a vat containing diluted pulp. A layer of wet pulp is formed on the cylinder. Water drains off on breaking the surface and is transferred to the underside of a moving felt belt. The belt is pressed on to the cylinder by rubber rollers. In the multi vat machine as many as eight cylinders may be used. The wet layers bind themselves to each other forming a single web of board. It is possible to have different constitution for each of the layers of board. e.g. the top and bottom layer may be a strong chemical pulp with some loading to provide a white opaque surface, the middle layer may be of poor strength such as gray waste water pulp. The principle of operation is generally simple, a rotating drum made of wire, picks up stuff either in the direct flow or in the counter flow machine. The mat formed on the wire covered mold is transferred to a felt carrier which moves it to the next cylinder mold. This process is repeated till a multiply board of several plies is obtained. The felt picks up the first ply of paper and the wet fiber mat picks up the next ply. It is possible to construct board with different layers of different physical properties over each other. Low grade waste paper pulp can be used as the stuff for the inner plies while good quality pulp can be used for the outer plies. Wet paper always adheres to the smoother surface and as both the surfaces of the felt and the fiber mat are smoother than the wire the transfer of paper is nearly complete. There are two distinct ways in which the cylinders operate. The contraflow was the standard technique till recently and the alternate method of uniflow is being used.

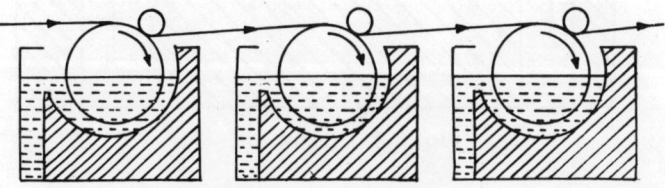

Fig. 5.1. Multivat board making.

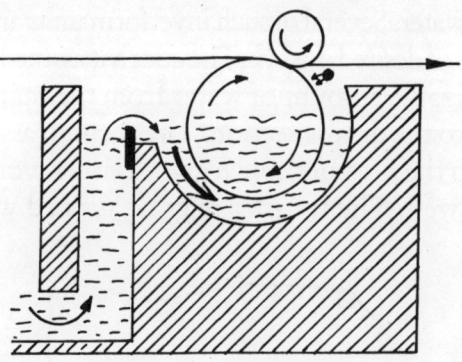

Fig. 5.2. Counterflow vat.

The drying follows the same pattern as in paper. However the major amount of water is sucked off at the earliest to reduce the amount of heat required. The drying section has a few calender stacks to provide a good surface finish. The inverform process introduced in the sixties was an important development in making paper boards. The stock is sandwiched between an upper and lower wire under a forming roll as soon as it leaves

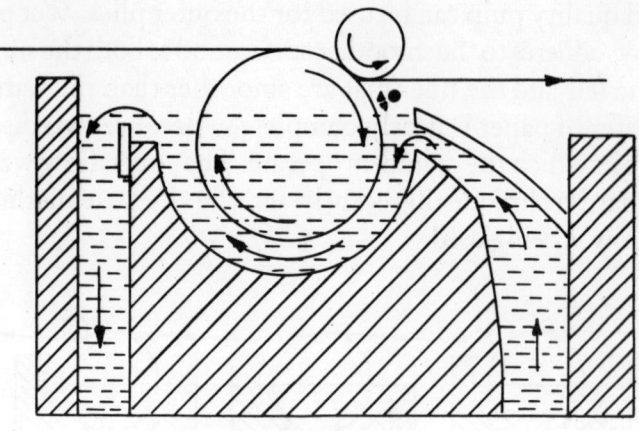

Fig. 5.3. Uniflow vat.

the head box. A stiff scraper blade held Inside the wire known as the auto slice scrapes off the water. Suction boxes take off more of the water. Several of such inverform units are combined to get boards of desired quality. The couch if assisted by vacuum removes a sizable amount of water from the mat before it is transferred to the drying section. The vacuum assisted couch compared to the pressure couch gives a larger working life to the wire as well. The transfer may be assisted with a small

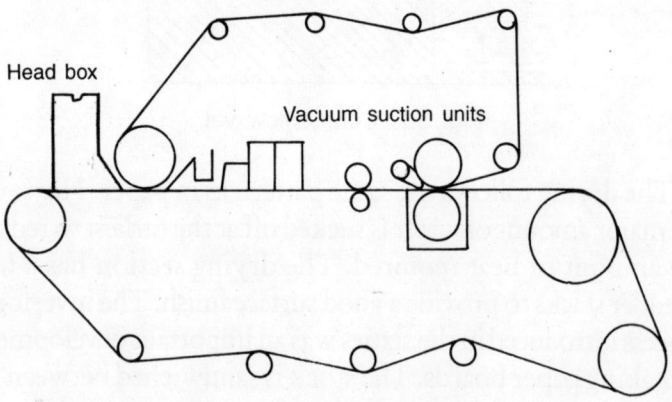

Fig. 5.4. Inverform double wire paper machine.

blast of compressed air or in some cases the drying unit may be placed very close to the wire section.

The inverform process introduced in the sixties was an important development in making paper boards. The stock is sandwiched between an upper and lower wire under a forming roll as soon as it leaves the head box. A stiff scraper blade held Inside the wire known as the auto slice scrapes off the water. Suction boxes take off more of the water. Several of such inverform units are combined to get boards of desired quality.

5.2 HAND-MADE PAPER

Making paper by hand was the only paper making process till the invention and adaptation of the paper machines during the 18th and 19th century. Paper making in India flourished in the Mughal period and gradually declined in its importance. Hand-made paper is made now only in small quantities compared to the machine processes. This is considered a luxury item. When properly handled, hand made paper has

1. An exquisite and elegant writing surface.
2. Has much better resistance to tear and has very good strength.
3. Has unmatched textures for the artist to use.
4. Can be made in many varieties and with many decorative inclusions.
5. Has high tensile, bursting, tearing strength.

Drawing paper for artworks, permanent document paper, card sheets, deckle edged stationery, unique carry bags, watermarked certificate papers, are but a few variations that can be made by hand.

The entire process starts with the preparation of the pulp. Generally unprinted paper wastes from printing units are used for paper making. However high grade deinked paper may be used. If available pulp from paper mills may also be used.

All fiber used for making paper must be free of mechanical and chemical contaminants and foreign matter. The pulp must also be homogeneous.

The main difference between the machine made paper and hand made paper is in the alignment of the fibers. In the continuous process of casting pulp to paper the fibers align themselves lengthwise in paper. The fiber bundles are difficult to break length wise but can easily be separated width wise. Hence paper properties vary strongly in the grain direction and the cross direction. In the hand made paper the grains are randomly oriented. In fact this randomness adds to the strength of paper in all directions.

Deckles and molds are made to the proper size. Usually the molds are square or rectangular. A mesh is attached to the wooden frame with panel pins along all the four sides. The mesh must be left taut.

Pulp is kept agitated in trough after ensuring that there are no lumps and the mixture is uniform. The frame is dipped into the trough full of pulp in proper dilution at an angle. When the frame is fully submerged it is kept horizontal and given a gentle shake sideways, forward and backward so that the fibers are evenly dispersed on the mesh. The frame is pulled slowly with shaking to ensure that no ripples are formed. The frame is held over the container for sometime to allow all the water to drain away.

The deckle is removed from the mold carefully. The paper now may be dried. The paper may be dried by leaving it on the mold for sufficiently long time or the paper may be transferred on to cloth for drying.

CHAPTER

6

Equipment Used in Paper Making Laboratory

6.1 PULP FREENESS TESTER

The freeness of pulps can be measured in a pulp freeness tester. The principle involved is to find the resistance offered to a mechanical handling in a dynamic condition.

Pulp is taken in a well cleaned calibrated container. The lids are sealed and the calibrated nozzle and screen plates put into commission.

Pulp is tested to the specifications as:

ISO 52672

CPPA C1

BS 6035/2

6.2 LABORATORY HAND SHEET PRESS

These systems let prepare hand made sheets from a given batch of pulp. The properties such as colour and tint, density and chemical properties can be reasonably projected after measurements on the hand sheet.

The sheets may be pressed as circles or rectangular sheets. The equipment may be pneumatically operated or may be manual. Heaters may be built into the units to estimate the drying time requirements.

6.3 LABORATORY CALENDERS

Generally built as 3 roller calenders—width typically 300 mm. Hydraulic cylinders permit application of large pressure—as high as 100 kg per linear centimeter. Heating with thermostat control is sometimes done.

6.4 LABORATORY COATERS

The equipment are built to conduct lab experiments on coating compositions and find the suitability on the substrate. Coating is applied by metering rods but blade coating attachments are also available.

The substrate is placed on a large horizontal table and coating applied under many varied conditions. The coating material may be heated prior to application. The table may be moved automatically. There are models that have built in drying chambers or infra red heating units to make possible quick drying.

SECTION

Two

Properties and Testing of Paper

Paper and
Board Properties

Printability and print quality are two properties associated with paper but not well-defined. In offset or gravure the printability would mean a measure of the missing dots.

Printability is defined as the extent to which properties of the paper lend themselves for true reproduction of a copy. Printability is not measurable in numbers as an absolute value as can be done in case of colour or brightness of paper.

The property can be considered as a resultant of the pulp properties and the formation parameters. The properties due to pulp are the ones such as porosity, scattering, opacity and absorbency. The properties that change due to formation are the ones such as the bulking thickness, surface smoothness, filler content and machine based variables.

The resultant is seen as

1. Ink show through.
2. Image clarity.
3. Ink requirement.

The result of the above variants can be approximately called printability.

Print quality is the degree to which the properties of the constituents including paper combine to obtain the desired result.

The following table summarises the result of printing on a few surfaces with specific inks. To an extent this is directly related to the property of the ink as can be seen as the requirements of two different grades of ink. Generally it is possible to standardise on some type of ink and get a subjective and objective idea about printability.

Grade of stock	Type of ink	
	Black	Opaque Orange
Enamel	360	250
Litho Coated	300	180
Label	240	160
Dull Coated	205	130
Newsprint	150	140
Antique finish	135	115
Machine finish	180	140

The table above indicates the approximate area of paper of the mentioned stock covered by a fixed weight of the given grade of ink. This gives an oblique measure of ink mileage that depends on the paper and ink properties. The results are obtained using a sheetfed offset machine.

7.1 DESIRABLE PROPERTIES OF PAPER FOR VARIOUS APPLICATIONS

7.1.1 Printing papers

Printing is the action of successfully placing ink on paper by various mechanical techniques. The paper has to adapt itself to the mechanical strain of handling as well as the chemical strain of ink.

Ink on paper has an important behavior pattern. This concerns the ability of ink to set as solid on paper. Ink is a solution/colloid/dispersion of dyes/pigments in liquid vehicles. The vehicles may be water based or oil based. The process of drying is when there is a phase change in ink. The transferred ink turns solid by any one of the processes such as evaporation, absorption or fusion. These processes are indicated in the figure below.

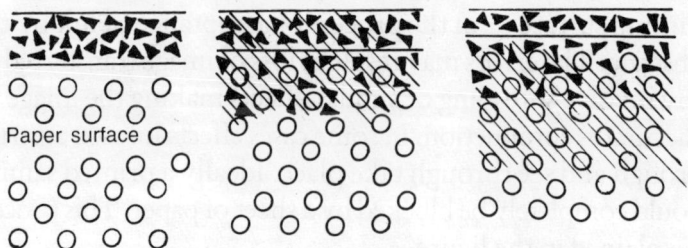

Fig. 7.1. Drying of ink on paper by absorption.

Ink is placed on paper in many ways. The most common procedure is to directly let the ink flow on to the substrate through small diameter tubes. The fountain pen or the quill are such examples.

Printing is transferring ink on to the paper or substrate surface with the help of a large printing surface—either directly or through a rubber blanket as in offset printing. Printing usually means direct transfer of ink on the surface by physical contact. In all the contact techniques, the ink is shared between the printing surface and the printed surface. There are chances that the printed surface may not accept ink due to certain problems of the surface. There are also techniques by which ink is squeezed on to the surface as in screen printing. The ink jet printing and similar processes place ink on the surface by non-contact techniques. In such cases the transfer of ink is direct and metered by the printing unit. All ink is either accepted or rejected by the printed surface.

The electro-transfer techniques depend on charging the printed surface and then expect the charged areas of paper to accept ink. The transfer is 100% but the ink need to be fused into the system. After transfer the ink undergoes the drying operation.

The ink may also dry by precipitation or reaction with the substrate.

There may be errors in the process of transfer from the printing surface or in the process of acceptance of ink by the substrate. The errors may result in loss of image transferred or may involve spreading or similar effects making the image an inaccurate reproduction. In some cases effects known as strike through and see through take place. Ideally a printed sample should completely be blocked by a sheet of paper. This process is explained in the figure.

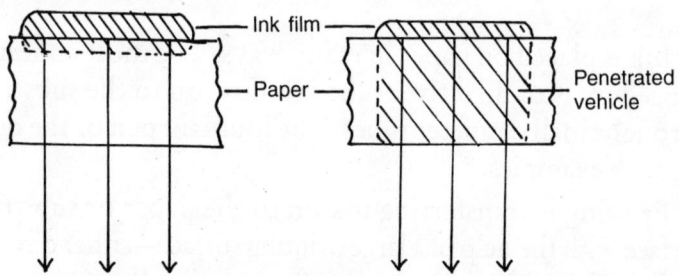

Fig. 7.2. Showthrough and strikethrough.

There is also the possibility that paper may be picked or torn by the ink which may be like a paste. The offset process in particular uses inks which are very tacky and is used in very thin films. In such cases the paper may stick to the blanket and when pulled may be damaged on the surface. In limits the coating may pull off or in uncoated papers the strands of fibers may be pulled off—contaminating the ink as well as producing an unsatisfactory print.

Ink dries on paper by any one or a combination of the following mechanisms:

1. **Absorption:** The pores in paper provide capillaries in which the liquid portion of the ink is absorbed. The vehicle tends to change its composition as it seeps through the thickness of paper. The insoluble constituents of ink are left on the surface. In rare cases the surface of dried ink may unbind itself from the ink vehicle and form loose powder. This also may lead to a dull printed surface. The effect is most common in moderate to low beaten paper.

2. **Evaporation:** The vehicle sometimes has many volatile compounds. When they are applied on paper or substrate that do not absorb, such as heavily coated or glazed papers, metals and non porous plastics ink loses its vehicle by evaporation. The rate of evaporation can be increased by application of moderate amounts of heat.

3. **Oxidation/Polymerization:** One of the major mechanisms of drying of ink is by polymerization of the constituents to hard masses. In linseed oil based vehicles the polymerization occurs due to oxidation of the ink in air.

4. **Chemical Polymerization:** In synthetic plastic inks the hardener may be a constituent of the ink itself.

5. **Precipitation:** One of the processes of drying uses inks that turn dry by precipitation. The precipitation is triggered with moisture and the mass turns dry.

6. **Fusion:** In case of toners transferred to paper or similar substrate the ink is in fine powder form either as a powder or finely dispersed in no polar solvents. The toner particles are deposited on the paper surface by electrostatic attraction and are thermally fused into paper.

7. **Dye Transfer:** Dye sublimation process is used in thermal printers to transfer ink on to paper. The process directly heats and sublimes the dye in its vehicle. Paper is supposed to receive the ink as vapour and immediately have it set as solid.

Evaporation	Oxidation/Polymerisation
Gravure and flexography on non-absorbent stock	Letter press and litho on non-absorbent stock
Gravure Flexography Letter press heatset Web offset heatset	Letter press Litho
Moisture set Steam set	News inks Book inks Non-heatset web inks
Precipitation	**Absorption**

Fig. 7.3. Printing process and process of drying.

The quality of the paper used and the ink selected should be such that there are no undesirable print effects and should not cause any problems while printing.

7.1.2 Process-wise demands

A listing of the process and the mechanism of drying commonly found is listed as below:

Printing Process	*Major Drying Mechanisms*
Letter Press	A, O, E, P
Litho Printing	O, A, C, E
Flexo, Gravure	E, A
Screen Printing	E, O, C
Dye Transfer	D
Laser Printing	F

Flexo: Ink is transferred from anilox rollers to the printing surface. The anilox roller cells have to be filled properly hence low tack ink is used. As a result picking of paper does not happen. Other surface properties are not very important. The ink generally dries by evaporation as solvent based inks are generally used. In cold countries heaters are used making it necessary to have good thermal stability for paper. Average mechanical properties needed.

Gravure: The ink is directly driven into the cells of the cylinder. The paper is pressed on. Hence pin point smoothness and absence of pits and voids is a critical requirement. Softness and compressibility are essential. Web should unwind smoothly.

Letter press: Smoothness is important for complete image transfer. Paper surface must be ink receptive. Water resistance not important. Pick strength important.

Screen printing: Dimensional stability and bulk are important requirements, others not very important.

Sheetfed offset: Higher surface and bonding strength is essential as inks used are very tacky. Water resistance is important as the paper would come in contact with moistened rubber blanket. Excessive moisture might introduce curl and dimensional changes. Long grain paper is required.

Web offset: The properties required are identical to sheetfed. The paper must have better tear resistance than as required by other processes. Ability to maintain uniform tension must exist to avoid wrinkles.

7.1.3 Writing papers

It is necessary that all writing papers must be well sized to avoid spreading of ink. The surface smoothness is also of importance. Papers moderately calendered but of good whiteness are generally used. In special applications papers such as vegetable parchment are used. Specially cast papers such as onion skin finish may be used.

The paper must be well sized as the ink should not spread and also should dry as fast as possible. The inks used are mostly liquid inks and often water based hence the choice is very critical. The surface should also be reasonably smooth and show through should be at its lowest.

7.1.4 Packaging papers

General printability, moisture and liquid resistance important. Must have good burst strength which is an approximate measure of the load bearing capacity. Must have good printability at least on one side. When white, must be bright enough to accept colours for the product. Folding strength is also important as packages may have to have some creasing and folding requirements.

Papers for corrugated packages must have good mechanical strength and adhesive acceptance properties. The outer side must have water resistance properties as defined for standard glazed surfaces.

7.1.5 Electrical applications

Paper is a good electrical insulator. Specially made papers are used as electrical insulators in capacitors, transformers, motors and chokes. In such applications the electrical properties turn more important than others. The paper is usually immersed in mineral oil. Hence oil absorption should not affect the properties of paper especially the mechanical strength. As oil absorption properties are difficult to measure directly, water absorption and retention is used as the guide. Printability and similar properties are not important but the surface properties such as finish and smoothness are known to improve the electrical properties.

7.2 PAPERS FOR OTHER APPLICATIONS

Paper for ink jet printing has desirable characteristics very similar to the writing papers. However the inks used are very thin and the major process of drying is absorption. Precipitation has also been tried as a drying technique in ink jet printing when the paper is specially treated or manufactured to set instantaneously. The paper need not have surfaces as smooth as for writing as the printing is done without direct contact.

Paper for xerox printing and copying also need be special in that the process of drying is melting in. The paper sheet has to stand fusion at higher temperature, which requires that the fillers or coatings must be resistant to heat.

Paper for packaging food articles may have to have some very special properties. These include toxicity, smell, bleeding of colours, grease and water resistance. Some of these properties are not quite measurable but are of great importance both as requirements of law and hygiene.

The physical properties measured generally indicate the physical condition of the paper sample. To quite an extent this indicates the printability of paper as the condition of the surface and the bulk are defined in these variables. The print quality depends on these properties to quite an extent.

The various properties of paper can be put into a few categories such as:

- Bulk properties
- Surface properties
- Optical properties
- Mechanical properties, and
- Others.

7.2.1 Bulk properties

Bulk properties are mainly a set of properties acquired in selection of pulp and fillers/additives. They correspond mainly to the density of paper and to an extent the thickness of paper. The thickness of paper is a manufactured property and in most of the applications has to be reasonable. The bulk of paper is specific to the application. The selection of paper or board is decided entirely by the application. For the same grammage of paper one may have different thicknesses. A reasonable thickness or bulk gives a sense of good touch and feel in a book compared to very thin pages. SGW papers are generally have a higher bulk. Dictionary and encyclopedia like large volumes of books have low bulk paper.

7.2.2 Surface properties

The surface properties especially of uncoated papers may have lot of variations. The mechanism of drying, the ink requirement and many such properties depend entirely on the surface. Calendering and coatings affect the surface properties quite considerably. The manufacturing method employed in making paper decides to some extent the surface properties. The surface properties are an important parameter in selection from the point of view of printability. Selection of the printing process may also depend on the surface available. The surface properties of coated papers are decided almost entirely by the properties of the coating material.

7.2.3 Optical properties

Optical properties are of two categories. The bulk optical properties such as colour, transparency/opacity (see-through) and the surface properties such as gloss, reflectance etc. Depending on the application one may have to choose the right values. The properties are a function of the beating, choice of pulp, additives and surface treatment.

7.2.4 Mechanical properties

Mechanical properties are a function of the formation and selection of fiber. They are not affected by normal surface treatments but are affected by lamination and similar processes. They are of great importance in packaging applications but are significant in choosing the right printing process.

It is possible to consider the mechanical properties in two different ways. The first is the mechanical properties of the thickness of paper and other the length and breadth of paper. The properties of the thickness are the binding of paper, variation in property because of the variations in composition over thickness such as two sidedness and fluff on surface. The length and breadth properties are the tear and pull strength, burst strength and properties related to grain direction.

7.2.5 Other properties

Properties such as moisture resistance, porosity, smell, taste, resistance to heat etc. are important in specific requirements.

8

Testing of Paper

8.1 SUBJECTIVE TESTS

1. Two sidedness

Two sidedness of paper is caused by loss of fine fibers from the wire side (hence leaving the wire side less dense and more porous) and slower settling of finer fiber due to larger hydrodynamic resistance. The wire side generally is rougher than the felt side. The water mark when created from wired dandy rolls is known as wove water mark. The paper when immersed in water or dilute hydrochloric acid decalenders and hence the twosidedness is more easily observed. In filled paper when marked with a coin the wire side has the lighter mark. Two samples of paper are torn from the sample, one with wire side up and the other felt side up. When compared a more feathered tear is seen on the wire side. Postage stamps may be printed on the wire side and gummed on the felt side so that the gum may be smoothly applied. Two sidedness may be even more critical in case of coloured papers.

2. Grain direction

An estimate of the grain direction and if calendering has been done can easily be found by visual inspection. If two identically wide and long pieces of paper are held together the one that sags less indicates that the grain direction is over the length. If a piece of paper is dropped into water the piece will tend to curl into a tube in the grain direction.

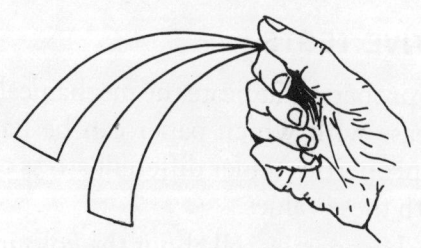

Fig. 8.1. Finding grain direction—the simple way.

A visual test against light will tell if the paper is closely made or not. Keeping a blank sheet over printed matter gives an indication of the opacity. By running a fountain pen over the paper it is possible to see if the ink penetrates through or the marking feathers when the paper may be slightly or under sized.

A piece of torn paper will give an indication of the tearing strength which is higher across the machine direction. The torn edge indicates the fiber length, the presence of multiple layers etc.

Grain direction can also be observed by

1. Visual inspection
2. Moistening and observing the curl direction.
3. Noting the way the burst occurs in burst strength measurement
4. Measuring the tensile strength in perpendicular directions.

Paper sheet when shaken will rattle if it is strong, well sized paper. If coated papers are rubbed together some white paper may come out. It generally is possible to mark coated paper with a coin.

The paper may have certain other physical properties which can not be quite measured. These include taste and smell of paper. They are of great importance in papers for packaging food.

8.2 OBJECTIVE TESTS

The strength properties indicate the mechanical properties of paper. The ease with which paper can be handled on the machine, its properties under different stress conditions are measured with these values.

The optical properties tell about the colour, opacity and gloss of paper sample. These are of great importance to get good print quality.

Chemical properties are a measure of the life of paper and also its suitability for different applications in terms of long term standing.

8.2.1 Physical properties

1. Grammage

Sometimes is referred to as substance. Thickness of paper expressed in terms of grams per square meter. A fixed area of paper piece is cut out from the given sample. Weighed in specially calibrated single pan balance with direct indication of grammage on scale.

A system of expression as basis weight instead of grammage is used sometimes. Basis weight is expressed as pounds per ream of paper, or pounds per thousand sheets or pounds per 1000 square feet.

The basis weight for some writing paper of size 17″ × 22″

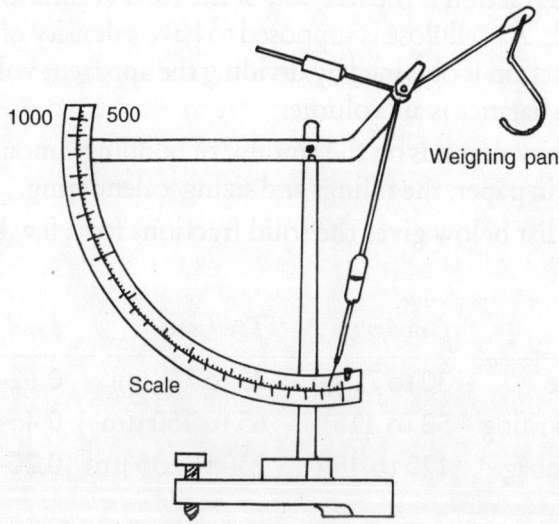

1000 500

Weighing pan

Scale

Fig. 8.2. A simple one pan balance to find grammage.

may be expressed as a certain number of pounds for 500 sheets e.g. 13 lb, 16 lb, or 20 lb. It is easily converted into metric grammage as 48.9 gsm, 60.1 gsm and 75.1 gsm.

2. Thickness

Thickness measured on single sheet when under a load of 50 kPa (1/2 atmosphere) in US and 100 kPa in Europe. Measured with precalibrated dial micrometer.

Thickness of papers is expressed in terms of points.

3. Density

Density is expressed as grams per cc. The gsm divided by the thickness in um gives the so called apparent density. This density is referred as apparent as it includes the air spaces within paper. Paper has a density of 0.5 to 0.8 though cellulose has density of 1.5.

Specific volume is the reciprocal of apparent density and is the volume in cc occupied by a gram of paper.

Solid fraction is the fraction of the total volume occupied by solids. As cellulose is supposed to have a density of 1.5, the solid fraction is obtained by dividing the apparent volume by 1.5 The balance is air volume.

Density depends on the amount of bonding, amount of air left out in paper, the fillings and sizing, calendering.

The list below gives the solid fractions for a few kinds of paper

Type	Grammage	Thickness	Solid fraction
Glassine	30 to 75 gsm	25 to 70 μm	0.62-0.75
Bond, writing	50 to 115	65 to 150 μm	0.45-0.65
Cement bag	120 to 195	230 to 635 μm	0.20-0.45

4. Porosity

This is an important property which decides the amount of ink vehicle that would be absorbed into paper while writing or printing. Besides the mechanical strength also depends to quite an extent on porosity. In applications such as filter papers porosity is the property most used. It is generally measured by measuring the leaking time of a known volume of air through

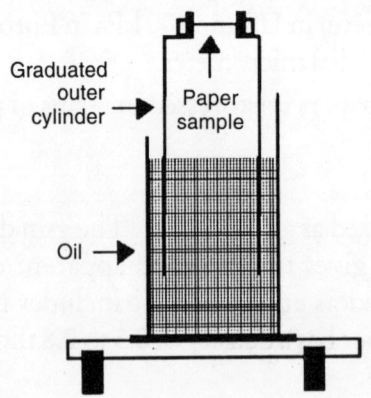

Graduated outer cylinder →

Paper sample

Oil →

Fig. 8.3. Testing porosity by leaking air.

a known aperture covered with the paper under test. The pressure differential is maintained as small as possible because the leak rate may depend on the pressure of contained air.

Porosity is measured as the time needed for leaking 100 ml of air through an area of 6.45 cm^2 of the paper or board. The method is not suitable for corrugated boards. Gurley instrument is used in measuring porosity. The sample may be clamped on top with screws or with a latch at the bottom of a metal cylinder weighing 567 ± 0.5 gm with a internal diameter of 74.1 mm. This produces a pressure equal to 12.4 cm of water. This cylinder is placed in an outer cylinder filled with lubricating oil with a viscosity in the range of 10 to 13 cPa. The sample is mounted with gaskets so that an area equal to 28.6 mm diameter is subjected to air passage. The time taken by the cylinder to move from the first graduation to the next of 50 ml is noted correct to 0.2 sec. The porosity is reported as the average time in seconds required to displace air of volume 100 ml through an area of 6.45 cm^2.

Porosity is of great importance in filter papers and air mask filters. Straw pulps mixed with coniferous wood pulps produce better filter papers. Asbestos fibers may be added to change the properties of filter paper.

Void volume is of great importance in insulating papers.

5. Smoothness

Smoothness decides many of the properties of printability including the consumption of ink. Coated papers are generally smooth and hence may have better printability for halftones. The pressure required in transfer of ink depends on the surface as it is difficult to fill the valleys with ink. Speckle (nonprinting by certain cells) a specific problem in gravure is directly related to surface smoothness. There are many techniques for measuring the smoothness.

Examining paper under a low power microscope can give a

reasonable idea about smoothness. The relief effect can be enhanced by illuminating paper at low angle. This technique reveals besides other things loose fibers, coating potholes etc.

The micro contour test involves application of a wax like thick ink onto paper and doctoring off the coating. The pigmented portion still entrapped gives an idea of the smoothness.

The Benderson test measures the rate at which a given volume of air leaks through the gap between paper and a metal ring. The pressure is maintained very slightly above atmosphere. Smoother the surface more tight the sealing and hence a larger time to leak. There are many similar equipment which try to overcome the problems if any of the above procedure. The print surf tester (using a sophisticated blow and flow measurement system for finding leakage) and the chapman smoothness tester (an optical method where a glass prism is placed on paper and the light reflected from the paper as a whole is compared with the light from under the glass—assumes that glass smoothens out paper) are other ways of checking the smoothness.

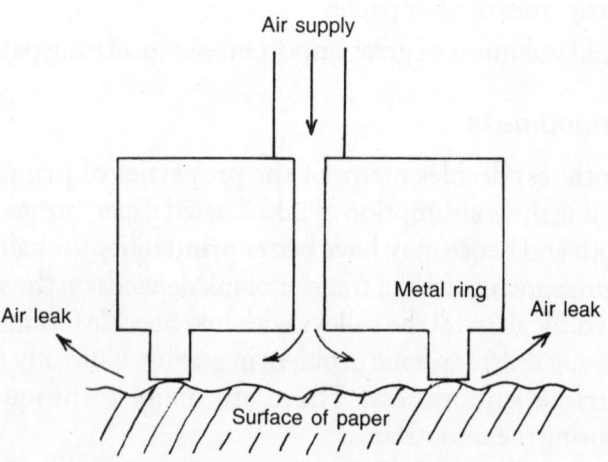

Fig. 8.4. Benderson smoothness tester.

Taylsurf method a process of amplifying and measuring the movement of a fine stylus on the test surface is also used often only for comparing two brands of paper. The surface is magnified and indicated in charts.

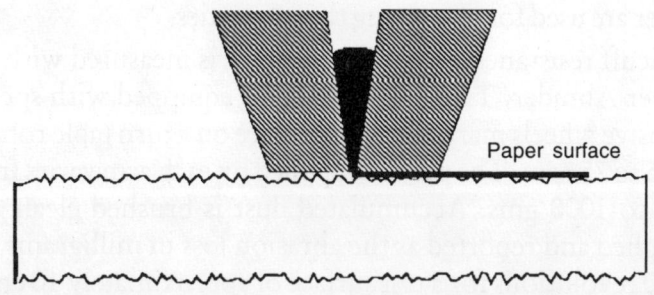

Paper surface

Fig. 8.5. Finding the larger volume of oil needed on uneven surface.

PIRA oil track method looks for the mileage of a given amount of oil as transferred to a surface and equates this number to the smoothness of paper.

Smoothness is enhanced by higher amount of beating, shake on the fourdrinier, increased wet pressing and increased calendering, surface sizing and pigment coating.

Smoothness improves the appearance of paper, it improves the movement of the writing instrument. It is indicated by the average size of surface irregularities.

6. Softness, hardness, compressibility

Softness is the absence of hardness when paper is crumpled with hand. Density, rigidity, compressibility and surface smoothness decide the softness of paper. An empirical method of measuring softness is to measure the volume under a fixed load. Softness is an important attribute of cosmetic papers.

Hardness is the tendency to resist indentation by other materials. The term hardness is used in identifying the amount of delignification that has taken place. A well purified pulp will produce a soft fluffy paper and a hard pulp a comparatively harder paper. Compressibility is the change in thickness produced by the application of a static load.

The Bekk hardness tester, the Gurley-Hill softness (SPS) test (a static method of measuring compressibility) and the CR tester are used for measuring the properties.

Scuff resistance or surface hardness is measured with the Taber Abrader. Two weighted arms equipped with special abrasive wheels mark the sample place on a turn table rotated at 65 to 75 rpm. The pressure applied by each arm varies from 500 to 1000 gms. Accumulated dust is brushed clean and weighed and reported as the abrasion loss in milligrams per 1000 revolutions for a test surface of approximately 10 cm^2.

Abrasion resistance is affected strongly by high moisture content.

7. Dimensional stability

The physical size of paper is affected very strongly by the moisture content in paper as well as in the atmosphere. All papers expand with increased moisture and contract with reduced moisture. At the time of manufacture paper is dried under tension. The paper in storage hence has tendency to shorten in its length. There is a marked hysteresis in the change of dimensions.

The dimensional change is anisotropic on paper. The ratio of cross direction and machine direction change is always greater than 2 to 1. The stability figures change from the edge to center of paper.

Treatment of paper to reduce dimensional changes:

1. Using non-cellulosic fibers such as glass.
2. Running bulkier sheets.
3. Reduce internal bonding by using stronger fibers, reduced beating and by adding fillers.

4. Add a bonding agent not susceptible to moisture such as urea formaldehyde resin.
5. Coat the paper to make it impervious to water.

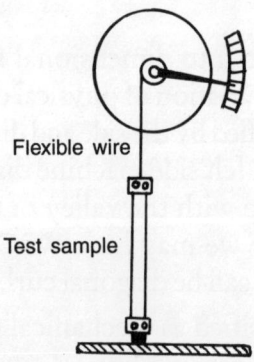

Flexible wire

Test sample

Fig. 8.6. Measuring paper variables.

One simple way of assessing the dimensional stability is by measuring the change in length of a strip of paper. A strip of paper is clamped between a fixed point and a pulley, weight and pointer mechanism to indicate change in length. The scale and pointer magnify the expansion.

Generally the printer wants to know if the paper is in balance with the machine room conditions or not. This is done by the PIRA Paper Equilibrium Tester. A sword shaped bar has two needles at a distance of 20 inches. The sword is thrust into a stack of paper and the pins activated to produce pair of marks. This marked sheet of paper is removed and balanced with the press room conditions. The paper is laid on a flat table and the

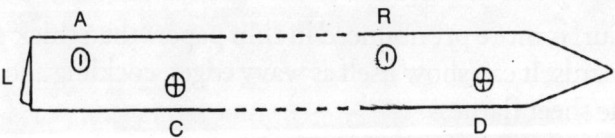

Fig. 8.7. Paper equilibrium tester.

marks compared with a glass graticule where a pair of markings at the same distance are compared. An expansion greater than 0.5% is likely to affect the side of a stack of paper.

8. Curl

Curl is closely related to dimensional stability and can be considered as local variation of physical dimensions in a large area sheet. It is identified by the side and direction. For example **FSMD** would mean felt side machine direction—paper curls towards the felt side with the valley of the curl in machine direction. Similarly we may have **WSCD**—wire side, cross direction etc. There can be diagonal curl.

Curl can be classified as mechanical curl, structural curl and moisture curl. Mechanical curl is caused by subjecting the sheet to mechanical strains beyond the stretching limit. The mechanical curls are often easily removed by subjecting the paper to a reverse curl. Certain kinds of curls caused by prolonged tension lead to micellar creep which may not be reversible. Paper roll stored that way for a long time is likely to develop this kind of curl.

Most of the curls are structural curls built into the interior or parts of the surface paper during formation and drying. If strain is greater on one side than the the other the paper will curl. At the time of calendering a difference in moistness of the two sides also leads to a curl.

Moisture curl is caused because of larger number of fiber crossings on one side than the other. At a given atmospheric moisture level the paper may be very flat but at a different humidity curl can occur.

Curl is more pronounced in thin papers than thick papers or boards. It can show itself as wavy edges, cockling and bends in the sheet flatness.

Curl is not quite measurable than simple visual estimates.

It must be avoided for true feeding of paper and hence good print.

8.2.2 Strength properties of paper

1. Tensile strength

Tensile strength is the load which has to be applied to the ends of paper for it to break. It is measured as the force required to break a given piece of paper 15 mm in width and 180 mm in length, that is clamped at both ends. It is expressed in kgf.

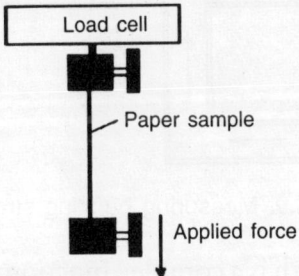

Fig. 8.8. Measuring tensile strength of paper.

Bekk or Vander Korput instruments apply a sudden load to the test strip and measure the work done in breaking it. Load cells are used to find the force of tension.

Tensile strength of paper is always greater in the machine direction. This is because of better alignment of fibers. There appears to be a linear relation between the area of the fibers in a cross section and the tensile strength. Paper has a strength in the region of 0.5-8.0 kg/mm^2 compared to about 80 kg/mm^2 for viscose. Tensile strength may be proportional to the square root of the fiber length.

2. Bursting strength

Bursting strength is the maximum pressure that can be sustained by a circular piece of paper diameter 30 mm, expressed in kg/cm^2.

In the Mullen burst tester the paper sample firmly clamped by a ring 6 inches diameter and hydraulic pressure is applied behind a backing diaphragm by pumping glycerine or ethylene glycol. A gauge records the pressure at the moment of burst.

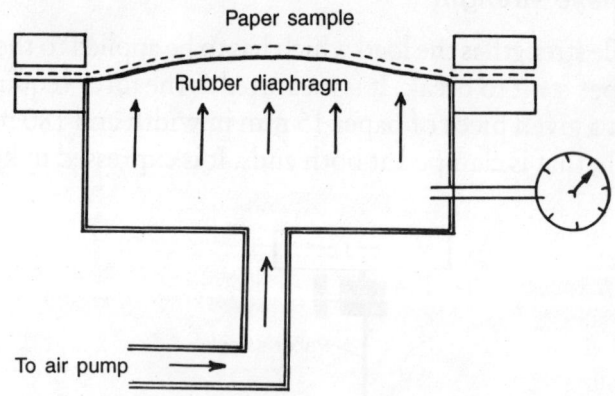

Fig. 8.9. Measuring bursting strength.

Bursting strength is a complex function of tensile strength and stretch. The stress exerted is mainly tension at the point of burst. It is possible to modify bursting strength and tensile strength by drying the paper under tension.

3. Internal tearing resistance of paper

The equivalent force required to bring in tearing off of a precut 50 mm wide piece of strip from a larger piece with a weighted pendulum. The test may be performed on multiple sheets.

Elmendorf tear tester is generally used. Test samples of standard dimensions are prepared on a special guillotine. The force of resistance offered by a cut strip from the sample is recorded.

The Elemendorf value does not indicate the resistance offered by a straight edge but considers the strength of a tear already started. The edge tearing strength is measured with Concora torsion tear tester.

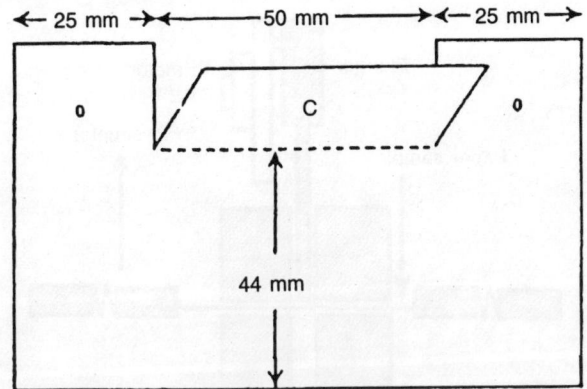

Fig. 8.10. Paper cut for Elmendorf test for tear resistance.

The tearing resistance depends on 3 factors—total number of fibers participating in the sheet rupture, length of fiber and number and strength of fiber to fiber bonds. The number participating is decided by the grammage of paper and the stiffness of the paper. Very rigid sheet would mean a much smaller number than a flexible sheet as the force will be distributed over a larger number of fibers. The force required to tear is much less than the force required to break. The force is used in tearing off some fibers as they are from the body of the paper and breaking them physically. Beating increases the cohesion and increases the forces of bonding between a given number of fibers. To an extent beating increases the number of fiber to fiber bonds adding to tearing resistance.

4. Folding endurance

A piece of paper if repeatedly subjected to folding under tension it will eventually break. Assuming a double fold to be one cycle a sample of paper 15 mm wide is folded many times. The average of machine direction and cross direction folding cycles is folding endurance index. Schopper folding endurance tester is used which forces double folds on a stiffly mounted test sample and measures the endurance in cycles.

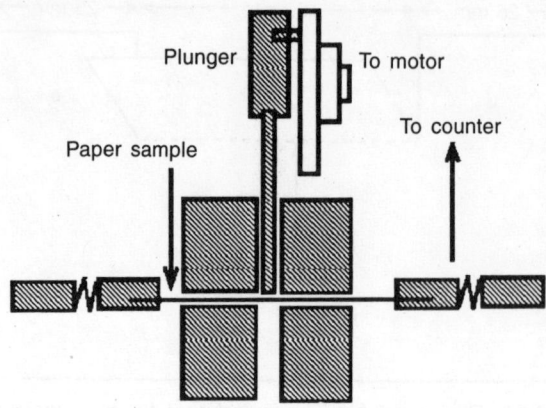

Fig. 8.11. Folding endurance tester.

It has been observed that in some cases the folding endurance in the cross direction is better than in the machine direction. This is in spite of having higher tensile strength in the machine direction. It has been surmised that some flow properties are involved in folding. Some theories indicate that the reason could be the lower strain in the cross direction. This stress relaxation may be the cause of higher creep. The folding properties to an extent depend on the grammage for a given composition, increasing proportionately with grammage. In grammages over about 100 gsm folding endurance starts dropping.

5. Stiffness

Stiffness can be considered the strength of paper in terms of bearing its own weight. It can be considered related to flow property.

In case of heavy paper boards static bending has been measured to indicate stiffness. Paper makers stiffness is a measure of the ability of paper to support its own weight and is proportional to Young's modulus (E), times the moment of inertia (I), divided by the grammage.

$$\text{Stiffness} = \frac{ET^3}{12} \times \frac{W}{L^2}$$

where T = thickness of sample, L = length of sample, and W = width of sample.

Stiffness is measured by the Gurley, Taber and other such instruments.

Stiffness of the paper increases as the cube of the thickness. It is also related to brittleness, rattle and other less definable paper properties.

6. Impact resistance and crush properties

With increased use of paper in packaging and as a structural material there is a greater interest in its impact and crush properties. TAPPI has a procedure for measuring the flat crush and ring crush property of corrugated board. Similar standards exist for drum, drop, compression and other tests for fiber board. These basically are a measure of the internal bonding strength.

7. Pick resistance

When printing by contact about half the ink is transferred onto paper which has to break itself away from the original thickness of film. This means a strain of separation exists which acts on the surface of paper that has received the ink. If the surface strength is poor or if the surface carries loose particles or fiber, ink does not get transferred. Ink picks up these loose fibers causing picking. This effect is going to be even more pronounced when the machines run at higher speeds and in multicoloured printing. Measurement of this strength hence is of great importance for all impact printing applications.

The pulling away of individual fibers from the surface is known as fluffing. The loose material may adhere to the printing surface or blanket causing errors in following prints. One of the ways of testing for fluff resistance is to run a print and

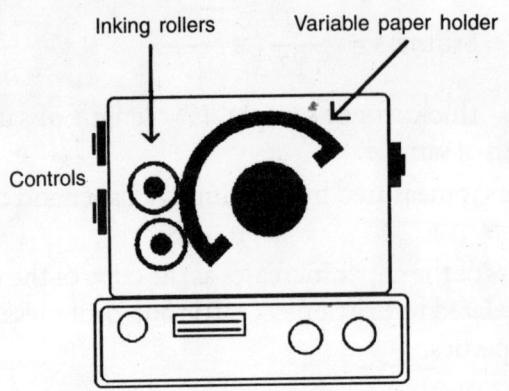

Fig. 8.12. IGT tester for printability and pick resistance.

observe the blanket for loose fibers carried. The PIRA procedure involves passing five sample sheets between two rollers one covered with a blanket that collects the fluff. The loose fibers are counted and reported as the number of fibers per sq. cm.

Lifting of areas of paper under printing conditions is known as picking. In rare cases paper or laminate may split under printing conditions. The best test is to go through the IGT (Instituut Voor Grafische Technik, Denmark) printability tester. Strips of paper 25 cm long and 3 cm wide are printed under a variety of printing pressures. A heavy pendulum falls and prints with increasing speed over the 150 degree arc of the impression sector. The first point where picking starts is identified and hence the suitability is identified. The method is excellent excepting that the equipment is very expensive.

Dennison wax test is an interesting simple substitute for finding the picking strength of paper. A set of eighteen blended wax sticks of varying grades of adhesiveness are available. Each stick is numbered and coloured for easy identification. The ends of the wax sticks are melted in a flame and pressed on to the sample paper. The wax is allowed to solidify and cool for about fifteen minutes and pulled off from the sample. The

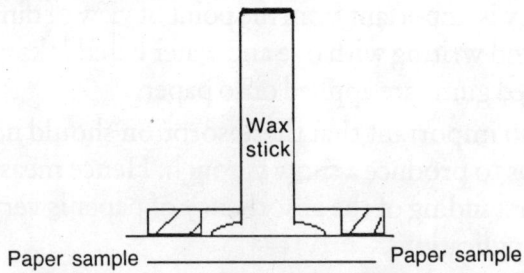

Fig. 8.13. Dennison wax test for pick.

flattened base is examined for traces of paper on them. Higher the dennison number greater the force exerted. The number of the weakest wax is representative of the picking resistance.

Though this test is not similar to printing, a number of at least 6 is required for reasonable printing.

8. Wet tensile strength

Wet tensile strength is a property which may be measured similar to dry tensile strength under different soaking conditions. This is important from the view of water resistance for packages and documents. The test is performed generally on kraft like packaging papers. It may be done on laminated papers too.

9. Absorbency—Water and oil

Paper as formed is a a very porous material. There are lot of air gaps and cavity in the body of the paper. At the time of manufacturing it is possible to change the consistency of paper to the totally absorbent antique paper—blotting paper to almost non absorbing glassine paper. There is a major difference in the way oil and water are absorbed by paper. Water chemically acts on the hydrogen bonds holding paper together being a polar liquid while oil being non polar acts entirely differently. Oil absorption is the most important factor in printing with inks which have to be partially absorbed while the water absorbency is important from the point of view of dimensional stability and writing with dye and water based inks and when water based gums are applied onto paper.

It is also important that the absorption should not be too much so as to produce a show through. Hence measurement and understanding of the absorbency of paper is very critical in paper applications.

The PIRA surface absorbency tester consists of a brass roller that is used as oil transfer medium. A drop of paraffin oil of known weight and viscosity is placed on the brass roller. The metal roller is moved over a blanket placed on an inclined plane to spread and split the oil. The roller picks oil from the blanket and transfers it onto a sample sheet of paper in the second revolution. The time elapsed between oil touching paper and 75% of the surface film is absorbed is noted. Waiting for 100% absorption is not preferred as the film behaves abnormally towards the end. The test is very sensitive to temperature and hence must be carried out at a constant temperature, say 20 degrees.

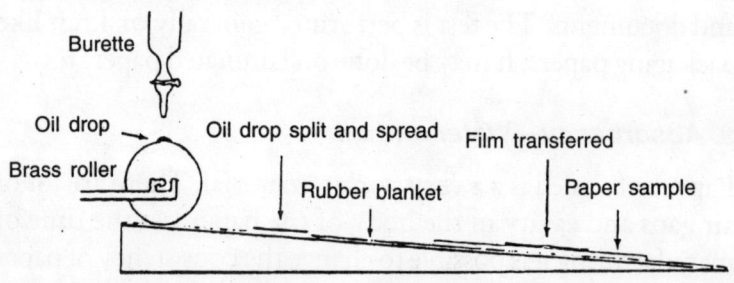

Fig. 8.14. PIRA surface absorption of oil test.

PIRA method of measuring oil absorbency and smoothness depends on the length of print that can be obtained by drawing slot of oil over the surface being tested. This is often done at different speeds so that the printing conditions could be simulated. IGT tester does the same thing in a different way. Dyed oil is used in both the applications.

A simple instrument known as the Cobb tester for water absorbency measures the amount of water absorbed. A ring of known inner area covers a weighted sample of paper. The area within the ring is soaked with water for a definite period of time say 60 seconds. The gain in weight is found out. The surface absorbency is given by

$$10,000 \times \frac{\text{Gain in weight in grams}}{\text{Circular area of ring in cm}^2}$$

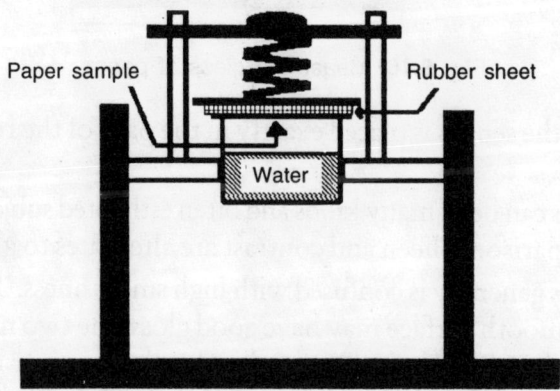

Fig. 8.15. Cobb surface absorbancy test equipment. The paper sample is held tight on to the container with a rubber sheet.

8.2.3 Optical properties

1. Gloss

The gloss is measured as the ratio of the amount of specular light reflected from the paper surface to that from a standard surface. The instrument for measuring specular reflection is first placed on a black polished tile (this reflects about 26% of incident light at 75 deg.) and the meter is set to 100. The tile is replaced with the paper sample and the light reflected is read as a measure of gloss. If the reading is 13% the gloss is 50%.

The gloss meter itself consists of a light source that focusses light on to the surface at 75 degrees (in some cases 85, 80, 60,

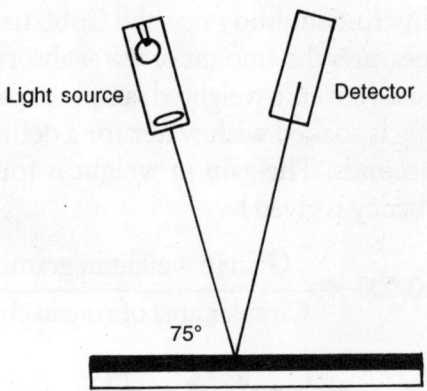

Fig. 8.16. Measuring gloss of paper.

45) and the sensor is placed exactly in the path of the reflected light.

Gloss can be of many kinds and often estimated subjectively by comparison. Sheen and contrast are alternates to gloss

Gloss generally is confused with high smoothness. Though a very smooth surface may have good gloss, the two never go hand in hand. Gloss can cause glare leading to poor reading quality. Mottle in printing leads to poor gloss, while good printed surface must be glossy. It is also believed that uniformity of gloss is more important than individual glossiness of ink and paper.

2. Whiteness and brightness testing

Whiteness is the extent to which a sample paper diffusely reflects all wave lengths. Reflectance is measured against a standard white substance such as Barium Sulphate. In the paper industry brightness is the percent reflectance of the blue light (wavelength centered on 457 nm) only.

Brightness is the overall reflectivity or lightness of the paper sample. In case of paper pulp it is the general bleaching efficiency of the yellowishness and grayness of pulp. A small amount of blue dye or optical brightness are added to improve brightness of paper. The brightness results are erroneous on coloured papers as the overall brightness of various colours are not the same.

The measurement often involves comparison of reflected light from standard objects—paper pads or ceramic tiles.

3. Colour measurement

Colour measurement and representation by munsell and other colour models and the colorimetric representation of colour are dealt with separately.

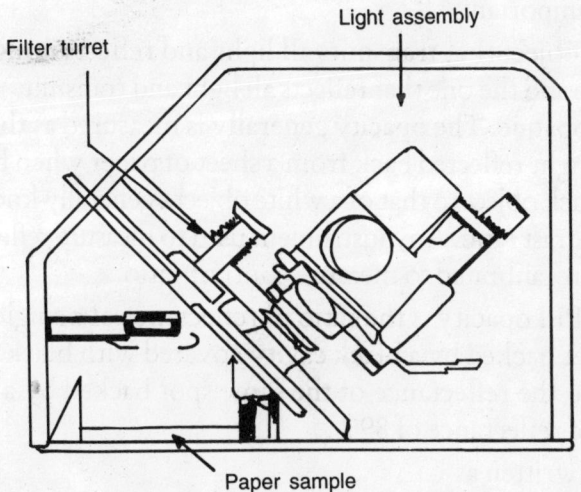

Filter turret

Light assembly

Paper sample

Fig. 8.17. EEL spectrophotometer—Measuring unit.

The colour of paper may be measured spectrometrically, abridged spectrometrically, colorimetrically and visually. The colour is also measured on paper as transmitted light and reflected light.

4. Opacity

Opacity is measured as the ratio of light blocked to light transmitted.

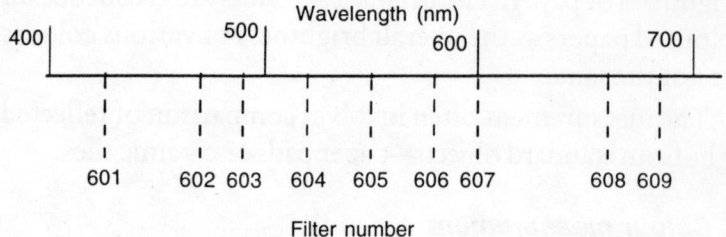

Fig. 8.18. Range of filters used in measuring colour with EEL spectrophotometer.

Transparency or the opposite of opacity is desirable in tracing papers. In most applications it is the ability to block that is important.

An object that transmits all light and reflects none is 0% opaque and the one that reflects all light and transmits none is 100% opaque. The opacity generally is measured as the ratio of the light reflected back from a sheet of paper when backed by a black object to that of a white object., generally known as the contrast ratio. Any instrument used to measure reflectance can be recalibrated to measure contrast ratio.

TAPPI opacity is the ratio of reflectance of a single sheet of paper backed by a black cavity covered with black velvet cloth to the reflectance of the same spot backed by a white body of reflectance of 89%.

It is written as

$$C_{0.89} = \frac{R_0}{R_{0.89}}$$

This method was used for quite sometime but now the white body is replaced with a stack of paper. This is known commonly as printing opacity as this simulates the show through that is estimated subjectively. It is not uncommon that the two opacities can be very different from each other and usually the printing opacity is a few percent higher.

Another type of opacity measured is the printed opacity.

This is referred sometimes to as strikethrough or showthrough. The test sample is printed on one side with a large solid black. The amount of light diffused back from the inked and uninked areas are compared.

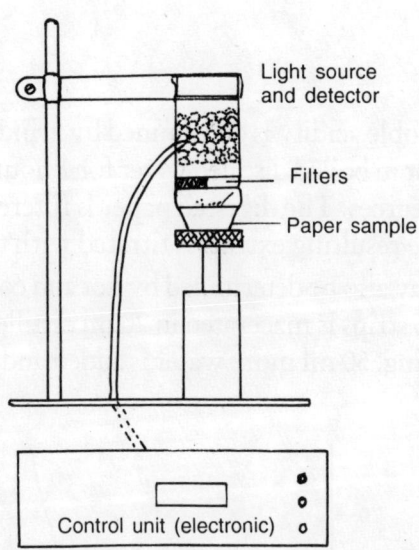

Fig. 8.19. A general colorimetric measurement instrument that is used for many optical measurements on paper.

Detailed measurements involving the scattering and absorption coefficients of paper in spectrometric terms are also done. (e.g. Kubelka, Munk equations)

The scattering power depends on the changes in refractive index, angles of incidence, other fillers in paper, the number of air pockets in between the fibers and the properties of the fiber etc. This is also dependent on the diameter of the fiber and the amount of fiber debris. The opacity is also changed by incorporating some other material in the space filled by air as in waxed papers. Surface coatings can also affect the opacity markedly.

8.2.4 Chemical properties

1. Taste and odor

Taste and odor are generally present in paper because of the process water. This must be eliminated if the paper has to have a good appeal and is of importance in food packaging applications.

2. Acidity

The water soluble acidity is determined by grinding the paper and digesting it in boiled distilled water for 1 hour by refluxing at 98 to 100 degrees. The digested paper is filtered in buchner funnel and the resulting extract is titrated with to a pH of 7.0

The pH may also be determined by hot and cold extraction. Paper cut into strips is macerated in 20 ml distilled water with constant stirring. 50 ml more water is added and the sample is

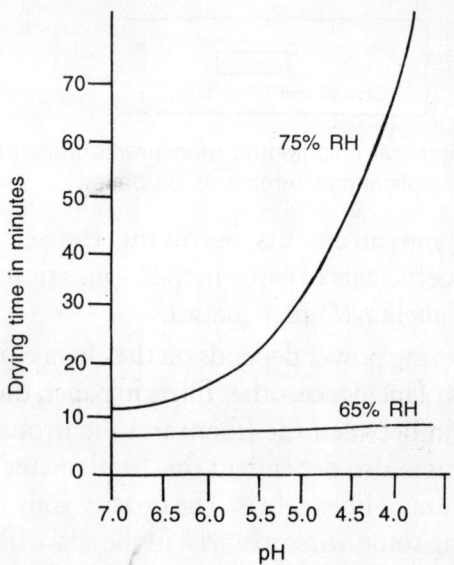

Fig. 8.20. Relation between drying time of ink and pH. Note its dependence on relative humidity.

left for 1 hour. If the maceration is done with water at 95 to 100 degrees it is known as hot extraction. Hot extraction generally gives a pH 0.5 lower than cold extraction.

Acidity in paper is the result of bleach residues left in the pulp, absorption of acidic vapours from atmosphere, presence of organic acids in pulp and coating materials.

pH is important for permanence of paper when it should not be lower than 5.0. Failure to permanence occurs by discoloration of paper, loss in strength of paper and change in chemical properties of paper. All these are caused by aging, due to the action of sun light, action of heat and moisture and mostly acidity. Ground wood papers generally are bright at a pH of 4.0 and that is the reason that wood based papers are not permanent.

The drying time taken by ink depends to quite an extent on the pH. This is indicated in the graph above. Samples of paper were studied at different pH levels for the time taken to dry a standard thickness of ink. The results clearly indicate the connection between the pH of paper and the time taken to dry.

3. Ash content

A weighed sample of paper about one gram in weight is ignited in crucible at about 850 degrees. The crucible is cooled in a desiccator and weighed. The ash often may contain the loadings if any used in paper.

Reducible sulphur in paper and other chemical test may also be performed.

4. Moisture content

Very dry paper is very hygroscopic. Moisture in paper is generally in balance with the atmospheric humidity. The moisture content increases drastically at higher humidity but is almost linear and about one sixth to 1 eighth in the range from 45 to 75.

A preweighed sample of paper sample is dried for long intervals of time at 102 to 105 degrees in air ovens. The loss of weight is recorded expressed as percentage of moisture by weight.

There are quite a few quick methods for measuring the moisture content. One of them for example heats the sample with an infra red lamp over a single pan balance and records the loss of weight till it goes steady.

Moisture content in paper varies from sample to sample and depends on two important factors. It obviously depends on the relative humidity. Larger the relative humidity greater the moisture content. It also depends on the physical condition of the paper meaning the manufacturing technique employed, the kind of fibers used and the kind of surface finish provided.

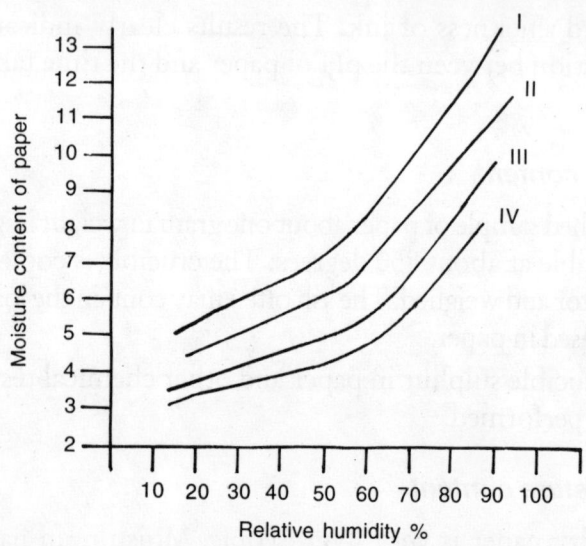

Fig. 8.21. Relationship between relative humidity and actual moisture content of various samples of paper. I. Manilla, II. SC mechanical, III. MF litho, IV. Bible tissue.

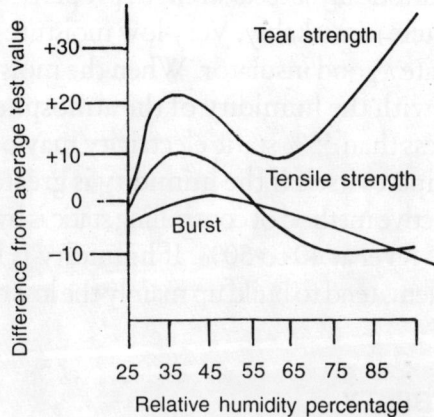

Fig. 8.22. Change in strength characteristics with relative humidity.

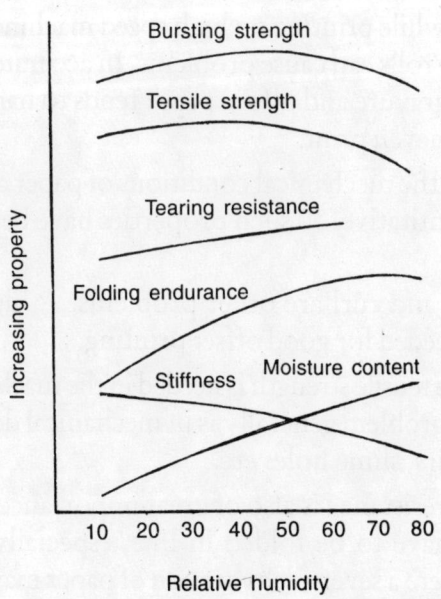

Fig. 8.23. Change in paper properties at different levels of relative humidity.

Moisture drastically affects strength properties and physical dimensions quite remarkably. Very low moisture leads to static as paper is quite a good insulator. When the moisture in paper is in balance with the humidity of the atmosphere and if the humidity is less than 35% static electricity may build up. This build up dissipates itself if the humidity is greater than 35%. The most effective method of controlling static is by maintaining the humidity level at 40 to 50%. If humidity is higher other kinds of problems tend to build up mainly the loss of mechanical strength.

8.3 RUNNABILITY

The runnability of paper falls into three categories—the mechanical, strength and moisture behavior conditions of paper. The sample must be satisfactory in all these categories.

The paper in rolls must be dimensionally true for easy unwinding while printing on high speed machines. Taper and ridges in the rolls can cause problems. In accurate nips such as the ones in gravure and offset, paper tends to hang at one end leading to uneven print.

Tests for the mechanical conditions of paper are subjective to semi quantitative. 30 such properties have been identified by TAPPI.

Fanning and curl are other problems. A higher level of flatness is needed for good offset printing.

Adequate tensile strength is needed in the machine direction though the problem generally is in mechanical defects such as cuts, felt hairs, slime holes etc.

Folding resistance is also of great importance especially if the forms have to be folded in line, especially in heat set printing where a severe dehydration of paper may happen.

Picking resistance or surface bonding resistance is also of great importance. IGT print tester may used for finding the values right. Picking may mean action of the ink with coating in addition to mechanical picking.

Behavior of paper under different moisture conditions is well discussed and the paper may have to be conditioned before printing.

Static is generated whenever two insulating surfaces rub against each other. Normally the moisture in the air discharges such static charge built up. Paper is a very good insulator. When the atmosphere is not very humid static starts building up. At humidity levels in the order of 45% and below this is very noticeable. Static electricity makes it difficult to separate sheets of paper. It may also affect ink transfer. Static must be eliminated by proper means such as grounding or discharging with ionizing lamps.

8.4 PAPER HANDLING AND STORAGE

A reel of 1000 mm in diameter and 1000 mm in width weighs differently for diffrent varieties of paper.

Paper Type	Weight approximate
Bulky	400 kg
M/F, Newsprint	500 kg
S/C Mechanical	800 kg
S/C Coated	900 kg

The reels are packed with end protectors and surface cover.

1000 sheets of demy sized sheets at about 60 gsm weigh approximately 26 kgs. Sheets invariably are packed and the weight of packing paper adds up.

Paper is combustible, easily affected by water and is easily torn and damaged. All transportation and storage has to keep this in mind.

When reels are damaged the damages are mostly on the

largest diameter and hence the loss is the largest. A cut 1 cm deep on the edge of a reel of 1000 mm diameter will lead to a loss of about 4% of paper in the reel.

If the edges of packets of sheets are damaged it will lead to problems of runnability. Often a second cut and hence an additional operation becomes necessary.

Adequate mechanical aids must be used for moving the heavy reels from the delivery bay to store and to the machines in a printing plant. Simple devices to turn the reels at right angles to movement must be used. The reels should not be damaged either by fall on to sharp objects or by piercing with sharp objects.

The reels are stored in some standard procedures. As far as possible the weight of paper should not affect the structure of the packings at the bottom when stacked. In case of sheets it is equally important that the packets should not stand on the edges and are laid flat. It is also important that the stacks should be left on even surfaces as damages and curl in paper are a usual problem due to such handling. This again lead to feeding problems on printing machines.

All the areas used for storage of paper must be water resistant. No seepage or flowing water should be there. If the store house is a part of a printing operation temperature and humidity must be made as nearly equal to printing conditions as possible.

Papers of different grades and colours must be well labeled and stored separately.

Fire protection must be provided in the storage area. Any or all of the systems i.e. fire alarm systems, sprinkler systems, fire extinguishers etc. may have to be used.

The storage area must be kept clean and free of oil slicks and the walls must be periodically checked for water leakage.

Applications

Varieties of Paper Grades and Their Properties

The properties of paper as explained earlier depends on the following factors:

- The composition of the pulp. The chemical constitution which in turn depends on the origin and the process of pulping.
- The physical condition and treatment of pulp. The amount of hydration and refining that takes place.
- Additives such as fillers and sizing agents and other pulps of different physical and chemical properties blended with the main pulp.
- Formation parameters. The amount of pressure applied. The design of the machine or the process employed in forming paper.
- Calendering or other mechanical treatment of surface of paper.
- Application of sizing agents or similar treatments to get desired chemical properties of the surface.
- Coating composition.

The various combinations of the above parameters lead to a variety of papers normally recognized by their trade names.

By using hardwood kraft pulp for providing bulk, with minimal beating and a slight hydration blotting papers are made. Soda pulp and cotton linters may be used. The product is bound to be bulky, porous and absorbent.

Antique paper is produced using hardwood kraft pulp for providing bulk. A small amount of sulphate and sulphite is used for better sheet formation. Sometimes made entirely from soda pulp

If a smaller amount of soda pulp is used and more of chemical fiber is included then the bulk is reduced. The beating may be increased to make papers of lesser bulk but of higher strength.

By increasing the beating and hydration further and using larger amounts of sulphite or sulphate pulp leads to a paper of better smoothness and firmness known popularly as English finish.

Increasing sulphate and sulphite and more of refining leads to a further better formed paper. This is strong enough and moisture resistant enough to be handled on litho machines and hence known as litho English finish paper.

Bond paper is produced using all sulphite or hard wood kraft pulp. The pulp is well beaten and well hydrated. The resultant paper will have a crackle and have a very high burst strength.

Coated body is produced with no regard for surface finish (as this would be obtained by coating any way) and hence use a high percentage of soda pulp. The sulfates may be added just enough to give the basic strength.

Coated or uncoated publication paper, tablet, litho, copying, business, cover stock, photographic, machine glazed and modified finish tissue, one-time carbon and other speciality grades are generally made using two and three press open draw fourdriniers. Plain, suction grooved, inverted, reversed and shrink sleeve presses are normally used alone or in combinations. Generally synthetic press felts must be used to reduce the breakages.

Kraft paper and bag, kraft liner board are made using softwood pulp which has long fibers. Machine speeds are similar to fine papers pick up felts must be open enough to resist filling. The second and third presses must have uniform pressure distribution to prevent base strike through.

Boards including food board, milk carton, cup, bristol, vellum, tag grades are manufactured on 2 and 3 press open draw fourdriniers. Bulk and good finish are the most desirable properties. Made from 3 plies of board, the liner, the filler and the back. The liner is usually coated with white fillings of good quality.

Newsprint and offset paper are lighter weight papers and must run at higher speeds. Stock for newsprint is primarily SGW and is characterized by short fibers and a percentage of long fiber chemical pulp is added for additional strength. Machines used are similar to fine papers. Pick up felts must be permeable and should not fill.

Single felt units are used in making tissue papers including toilet, facial napkin and toweling tissues. Yankee fourdrinier with two felts is traditionally used.

9.1 SOME SPECIALITY PAPERS AND BOARDS

Adhesive coated papers: Made from different paper bases such as glassine, coloured, glossy coated, cast coated, low gloss coated etc. Foil mounted on paper is also used. Gummed papers are more prone to curl and hence carefully handled. Adhesives could be heat seal, pressure sensitive and water based.

Blanks: Thick paper board with thickness 0.4 to 1.25 mm made on cylinder machines. It is designated by ply number.

Excessive bending and ink tack must be avoided. Used generally as backing material.

Cards and boards are generally made from chemical pulp. Grammage between 140 and 350 gsm. Because of the thickness it is possible to attain a variety of finishes. When rag content is increased a variety known as index bristol is produced which is ideal for index cards, erasable cards etc. Postal bristol is used for making post cards. Coated bristols are used in making base material good for picture post cards.

45 to 90 gsm paper is used in making business forms. The business forms are printed on web machines hence the paper must have good physical strength. Dimensional stability to stand collating and registration and computer print out is also an important criterion.

Papers for making carbon less papers have to stand the specific chemicals on the back of the top sheets and the front of the bottom sheet. Care is taken in printing so as not to damage the coated surfaces and not to use excessive pressures or folds.

Embossed papers made with engraved rollers.

Gummed paper: Paper with gum or adhesive coating on one side. The adhesives may be pressure sensitive or water soluble.

Hand-made paper: Stress free and grain direction free paper made on the deckle. Traditionally 3 finishes: (1) **Hot Pressed** (Zinc plates are placed between sheets and passed through rollers to give some glaze); (2) NOT meaning **not** hot pressed: When paper is dried by pressing 3 or 4 times between felts; and (3) **Rough** when only one pressing is done and the surface is fairly rough.

Kraft: The wrapping brown paper made from sulphate pulp has long strong fiber. MG is common often with a pattern on the glazing cylinder. Can be bleached white with difficulty and can also be made in colours.

Marbled paper: Used mainly as end papers of books. Original is always made by hand. Various liquid pigment colours float on a viscous gum solution and skilled craftsman weave patterns. Uncoated good quality strong paper is dipped sheet by sheet into the bath. Artificial marbled paper is produced by offset printing.

9.1.1 Coated papers

A light pigment coating of the order of 4 gsm may be applied as a preliminary coat. Sizing is a substitute for light coatings. The higher end is the cast coated board. Between the two there are a wide variety of coated versions. Blade coating produces a matt finish. Calendering is done to get a satin smooth finish. Kuster rolls placed at the finishing end of paper machine can improve the finish. Coating can be applied to all kind of paper or substrate.

Art paper: Heavily coated paper with high gloss. Three or so coats and super calendering done to obtain better finish. The first coat is usually on the paper machine itself. The next about 8 gsm may be off machine and blade/roller/air knife coated. Super calendering done to obtain desired finish. Expensive because of multiplicity of processes.

Bill blade: A technique of coating in which both sides of the paper are coated with about 10 gsm of coat material. The coating is applied with rubber backing roll and blade.

Blade coated, cast coated: Techniques used for coating. Both produce material with 80 to 400 gsm substance. Cast coating gives exceptionally high gloss.

Chromo paper: Generally one-sided art paper. Used specially for colour proofing. Made from good quality material and well finished and coated matt or bright white. If chromo has to be used for litho printing the coating as well the bulk must be reasonably water proof.

Light weight coated: Papers with substance of about 60 gsm. Used generally for journals.

Machine coated: Coating done on paper machine. Generally a cheaper alternative, hence used in educational books.

Matt coated art: Art paper may be matt or gloss finished. Used for expensive colour printed books. Easier to read as light does not reflect directly from surface. Glossy ink can produce excellent results.

Pigment coated, Roll coated, Size press coated.

Truflo: A high gloss paper with above average bulk. Super calenders are not used. Both sides of a two wire paper is coated to a gloss finish. 115 to 170 gsm is common.

Cover papers: Are specially made coated or uncoated for books (case bound or limp bound). The substance range is around 100 gsm. Grain direction is important. Book binding cloth is rarely used now and paper with textures is the common material being used. Sold by meters.

May be unglazed and fancy finished.

9.1.2 Boards

Coated boards (Art board): Used for covering paper backs and limp books. After printing the board is laminated substance range 200-500 gsm.

PVC: Incorporated into a base paper or applied as a coating produces a material between paper and cloth. 250 gsm is typical weight.

Uncoated paper between 150 and 500 gsm also could be used. Some are good as book end papers. Book Jackets Normally coated on one side only with strength durability and high gloss—around 120 gsm.

Substance over 220 gsm. Display boards and Grayboards (usually straw or waste paper).

Ivory board: Very smooth board for business cards, invitations and greetings.

Pulp Board uncoated wood free board—usually a single ply for paper back covers—need lamination.

9.1.3 Uncoated papers

Acid free paper: Uses chemical pulp pH 7.07 or more. Lignin is acidic hence it has to be lignin free.

Antique laid: Paper made with a laid dandy roll to get a pattern of noticeable parallel lines or chain lines. Other patterns also may be laid with dandy roll.

Antique wove: They are sold as 14 to 25 vol i.e. 14 or 25 mm for 200 pp at 100 gsm. Lower vols known as book woves. 70 gsm vol 21 paper will be less expensive than 80 gsm vol 19. Eucalyptus pulp (traditionally esparto pulp) which has opacity and bulk because of short fibers are pulped CTMP to get the grade. Lower vol have a smoother surface compared to higher vols. Up to vol 21 generally produces good quality prints. Novels, children books, fax editions etc. are produced with antique wove. The colour generally is cream. Are trough or matt finish.

Bank Note paper: Made by hand or mold. Selected linen cuttings and added ramie fibers used to get strength. Long and prolonged heating is necessary to get the most fold resistant paper.

Bank papers are rag papers normally tub sized. Thin and strong papers generally not glazed.

Bible printing: As it is thin may have problems of pin holes and see through. Folding and holding large sheets may be a problem. India paper was used by OUP. Smooth but not very polished surface. Generally all rag or rag and wood. Loaded with chalk to about 30% and a small amount of titanium dioxide.

Bleached mechanicals: Chemically bleached and some times super calendered. Light weight coateds can be used for newspapers.

Bond: Classified as writing papers. Must have permanence that is durability in handling, erasing, folding etc. Enough surface hardness for pen and ink writing, typing, stiffness as required for letter heads etc. Generally made from chemical wood.

Cotton content bond papers with cotton (used to be called rag content) between 25 to 100 p.c. They are made on slow paper machines with air drying and well formed papers. They have added strength and a cockle finish. Due to the hard surface the right ink must be used.

High quality papers are used for policies, deeds etc.

Laid finishes are also possible and may be in colour.

Weight range 60 to 80 gsm. Alum and rosin are common sizes. Engine, surface and tub sizing may be done. Very similar to Bank paper but is stronger and thicker.

Book woves: Low bulk antique woves.

Bulky MF Mechanical: This is pure mechanical or bleached mechanical with a semi smooth machine finish, ideal for paper backs. The fibers are shorter than chemical fibers and hence have a bulk and opacity. It has a creamy off white shade with a brightness between 70 and 75. The brightness enhanced by having an extra peroxide bleaching. Substance is between 40 to 80 gsm and thickness 70 to 125 microns. They are supplied in reels. Generally by length and sometimes by weight.

Bulky news: Same as above no sizing and lesser smoothness good for bad paper backs.

Cartridge: Weight between 90 and 220 gsm. Used mostly in litho printing. Most popular type of printed surface. Unbleached or semibleached made from manila, jute and kraft pulp.

Copier paper: Higher dimensional stability desired. Should

not curl or jam in copier should stand the heat in copier and should be made with duo formers. Substances 70-160 gsm.

Heat set web offset: This grade may be exposed to open flames and hence moisture content should be low—about 5%.

Light weight printing: Uncoated wood free MF used in dictionaries and the like.

Mechanical printings: Superior newsprint with engine or surface sizing.

MF: Used extensively for magazines and general printing work. Generally made from wood free and mechanical fibers and are surface sized. Finish depends on calendering.

Neutral sized Paper: Alum is acidic and if the loading is mostly calcium carbonate it may be possible to get paper with pH 7.5.

Newsprint: Brightness 60 ISO, discolours quickly made almost entirely from ground wood, may be calendered to print colours, screens over 100 not ok. Standard substance is 45 to 50 gsm.

Offset cartridge: Paper over 90 gsm specially made for litho printing. Particularly suited for children books and colour work. Surface sized, uncoated and are usually wood free. CTMP pulp may also be used. Paper is bright white may be acidic with pH 5 but must be very flat.

Offset printing: Weight less than 90 gsm but similar to the one above. Both wood free and mechanical are available and are classified as such. Used extensively in educational books etc. of longer life.

Opaque printing: Generally surface sized, lightly coated less than 90 gsm.

Part mechanical: With more than 10% mechanical pulp. (Not wood free) CTMP and TMP based pulp combinations are good for most of the jobs.

Permanent paper: Paper made from rags and cotton in

Europe and America have stood the test of time. They have remained almost unchanged even after four centuries. The only visible change is the colour which too changes only when paper is exposed to sunlight or ultra violet light. The degradation that occurs in paper over time is that it turns brittle and crumbles with time. Wood pulping got introduced into paper making in the later half of the nineteenth century. It was this paper which had a problem with its life. The problem was identified as the presence of rosin and alum as either sizing or digesting agents used in pulping. With reaction with air the materials yield an acidic ambient which degrades the bonding of paper. The problem persisted with chemical wood pulps and was sometimes even more severe. Since 1950 there is a clear tendency to move from acidic to either neutral or alkaline paper making processes.

The negative effects are more predominant in wood pulps than other sources of pulp such as rags. It is assumed that the mechanism of degradation is related to lignin and hence wood pulps free of lignin are known as wood free pulps. Paper counting less than 10% ground wood is considered wood free.(American National Standard for permanent paper—ANSI Z39 1984) pH 7.5, should contain an alkaline reserve such as Calcium Carbonate, a furnish of entirely wood pulp or ideally rag fibers, a specified tear resistance and a specified fold resistance.

Pure paper: Meaning contains less than 10% wood pulp.

Twin wire paper: Has much better formation on top. Depends on manufacturing process.

Water mark: Used from dandy roll as a security measure and also as a trademark. Cutting to sheets with register is difficult hence product more expensive.

Wood free: Paper made from fibers separated by chemical processes, free from lignin and other woody parts. May contain up to 10% wood fiber.

Web Sized Offset Paper: A cheaper substitute for the light

weight coated paper used in magazine applications—SC process used in making.

9.2 A FEW VARIETIES AND KINDS OF PAPER AND BOARDS

Bible printing	25-35 gsm
Lightweight Printing	35-60 gsm
Printings	60-90 gsm
Cartridges	90-220 gsm
Boards	Over 220 gsm

9.3 SUBJECTIVE TESTS IN GENERAL ON PAPER WHILE ACCEPTING FOR PRINTING

Estimation of:

- Capability of reproducing the image elements of the original
- Cause a minimum of printing problems.

The important parameters are:

- Print uniformity.
- Relative print density.
- Colour rendering.
- Ink requirement.
- Ink setting, set off, smearing.
- Print through, (show through and strike through)
- Rub off properties.

9.3.1 Newsprint

The only three tests done in a newspaper plant are:

1. Grammage: 52 gsm was a std. till 70. It has reduced to 49.8 gsm, 45 and 40 gsm and 36 and 28 gsm for air mail editions.
2. Moisture: Usually between 5 and 10%. About 1%

moisture is introduced in each printing unit. Less than 7% is not very acceptable.

3. Ash content: generally less than 2%. In German products may reach 10 to 12% due to fillers. Tolerance and acceptance limits are not specified.

Structural characteristics

- Felt marking and wire marking to be avoided for better quality print.
- Sheet density and thickness not specified.
- Surface smoothness is expressed as the ml of air escaping per minute with an air supply pressure of 1.471 kPa and at the edge 98 and 490 kPa.
- Smoothness should fall in the range 75 to 175 ml/min (Bendetsen 98 kPa). Higher values preferred by offset and lower by letter press.
- Bekk smoothness 45 to 75 seconds
- No general recommendation exists for compressibility, air permeability, hardness, oil absorption and water absorption.
- Shade values: As measured on STFI master Elrepho 2000 instrument. on 40 to 48 gsm.
- Y-Value 64.5%, Dominant wavelength—576.5 nm, Excitation purity 7.5%
- L* 84.2, a* -0.4, b* 6.9
- Opacity: 94 to 90%

9.3.2 Problems in web

1. Mechanical defects

- Edge damage.
- Body damage.
- Head damage: Damages to the side of the reel may be mechanical or due to moisture absorption.

- Out of roundness.
- Calender cuts: Cuts in the machine direction caused in calendering.
- Slime hole: A hole on paper produced by resin or fungus.
- Water drop hole: Hole in web caused by dropping water before complete paper formation
- Plucking hole: Produced by paper sticking to rollers on paper machine.
- Silver, shiver: Unseparated bundle of fibers in paper
- Hair cut: Cuts caused by thin foreign material such as hair.
- Winder wrinkle.
- Burst in reel.
- Uneven winding.
- Convex or concave winding.
- Run together at core.
- Loose paper.
- Edge cracks.
- Rough edges.
- Protruding splice.
- Stuck splice.
- Soft end, Baggy end.
- Rope marking.
- Core defects: Protruding, slipped, damaged.

2. Running defects—Runnability of webs

1. Web break at press start:

Before printing unit   Mechanical problems of early knife activation or baggy parts on roller.

Tension control not ok in or after printing unit.

	Too tacky ink.
	Water in cylinder gap.
	Improper start up sequence.
2. Web break during reel change	Improper glue.
3. Web break in run	Transport damage in reel.
	Too high winding tension.
	Too high tension on outer layers.
	Imperfect mill join.
	Out of roundness of reel.
	Holes in the web, shives in web.
	Too much too tacky ink.
	Water, ink or foreign object on web.
	Creases on the web.
	Wandering web.

9.3.3 Sheetfeed problems

- Double sheets—due to many problems.
- Sheets sticking and separation problems—hand jogging needed often.
- Sheet defects—incorrect edges, damage in transit and others leading to transport in machine problems.
- Mechanical strength problems—softness, tearing strength etc.
- Paper strength problems, paper sticking to roller and tearing.
- Compressibility and surface strength—marking by grippers, tear at grippers.
- Problems due to static electricity. Generally sticking together of paper sheets and hence double sheets.
- Uneven moisture on different areas due to many causes—

can be cured sometimes by leaving the packet of paper to adjust to room ambient.

- Tail end hook.
- Caused by the paper sticking to the blanket too tightly when pulled off. Can be remedied by using heavier paper or adjusting ink tack.
- Wavy edges are caused by stack of paper at lower moisture level than atmosphere.
- Tight edges reverse of above. The center of the sheets goes baggy.
- Creasing further aggravated because of water.

9.3.4 General problems in print reproduction due to paper

Printability:

- Paper colour, inability to print some colours.
- Gray or faint print, insufficient ink, ink/water imbalance.
- Colour variation throughout stopping and starting during run, ink/water imbalance.
- Shadows filling in on photos—print too dark, over absorbent paper, excess printing pressure, over exposure of plate.
- Unprinted spots on paper-picking, ink too tacky, too soft paper, wrong blanket.
- Spots on printed areas—hickies, paper fibers or dried ink blobs adhering to paper or blanket.
- Blurred colour photos-misregister, paper stretch due to many reasons.
- Surface smoothness problems.
- Loss of image—low spot, indents on blanket.
- Faint image outline marks on solid—repeat-ghosting, trapping, poor ink film adjustment.

- Smudge on non-image—scumming, plate corrosion, improper dampening.
- Faint tint on non-image areas—tinting, contamination of fountain solution.
- Faint pattern over print—damper marks, excessive pressure.
- Random image smudge—set off, excess ink, undrying ink.
- Distortion.

9.4 STANDARD PAPER SIZES

9.4.1 Standard sheet sizes of paper

- Crown: 15″ × 20″ at 8vo 5″ × 7-1/2″.
- Demy 17.5″ × 22.5″ at 8vo 5-5/8″ × 8-3/4″. Used as writing paper.
- Royal 20″ × 25″ at 8vo 6.25″ × 10″. Used as cover paper.
- Foolscap 13.5″ × 17.5″ at 8vo 4-3/8″ × 6-3/4″
- Medium 18″ × 23″
- Bond, writing, ledger, manifold 17″ × 22″

In British and American sizes the size is often referred as the basic size and the number of pages made from it e.g. crown octavo.

9.4.2 Metric sizes, also ISI

- A0 = 841 × 1190 mm (Area = 1 sq. meter—All types of textbooks, magazines, newspapers, periodicals, etc.)
- A1 = 594 × 841 mm
- A2 = 420 × 594 mm
- A3 are direct further half-divisions of A2
- B0 = 1414 × 1000 mm (Large size posters and calendars)
- C0 = 1297 × 917 mm (Envelops, small pocket size books, brochures, folders, etc.)

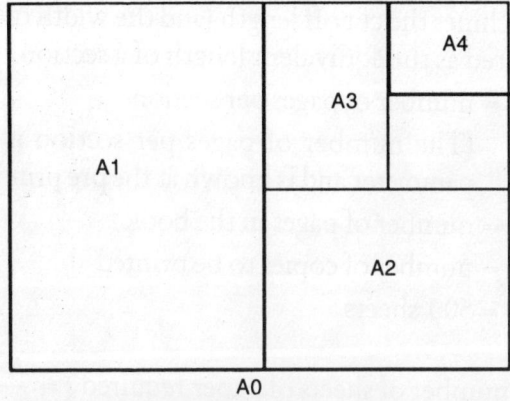

AO
Sub-divisions of AO paper

Fig. 9.1. Paper subdivisions in metric sizes.

Untrimmed sizes are made available which carry the prefix R and contain 5% more area of paper as trimming allowance. The division is identical as in AO.

- RA0 = 860 × 1220 mm
- RA1 = 610 × 860 mm
- RA2 = 430 × 610 mm.

The untrimmed stock is especially useful in creating publications whose pages are in the A range.

9.5 PAPER CALCULATIONS

9.5.1 Calculating the paper requirement for a typical publication to be printed

Section is defined as a part of a book that generally is produced from one sheet of paper. The standard size of paper selected is organized into the forms and the pages are arranged as per the imposition scheme. The paper wasted in between the page layout need not be taken into account for calculating the paper requirement as the cutting allowance as well as design wastage is accounted for. The number of pages in a section hence allow a direct calculation of the paper required for the book. In case of web machines the cut off length (and the width of web) can be considered as the equivalent length of a section.

p = number of pages per section

(The number of pages per section is a design parameter and is known at the pre printing stage)

n = number of pages in the book

c = number of copies to be printed

1 ream = 500 sheets

Hence number of sheets of paper required $s = \dfrac{n \times c}{p}$

Add to this number the wastage and the spoilage allowance. (This may be simplified as a multiplication factor such as 1.05—see end of this section for actual calculation. The spoilage allowance depends on the number of copies and the number of colours and also depends on the printing house.)

$S = s\,(1 + \text{spoilage allowance})$

Number of reams to be ordered = $S/500$

Example: To print 1000 copies of a book which contains 144 pages and made from 16 page sections arranged on 24″ × 32″ sheets:

The paper required is $s = \dfrac{152 \times 1000}{16} = 9500$

$S = s \times 1.05 = 9975$ (Spoilage assumed as 5%)

Number of reams of paper needed = $\dfrac{9975}{500} = 20$ reams

Calculating the weight of a ream of paper: If the paper is of any of the 'An' series of sizes then

the weight of 1 ream = $\dfrac{\text{gsm rating of paper}}{2 \times 2^n}$ kilograms

Example: The weight of a ream of 60 gsm paper that is in A2 size is

$$= \frac{60}{2 \times 2^2} \quad 15 \text{ kg}$$

If the paper is untrimmed a 5% extra has to be added. For example, the weight of a ream of RA0 sheets of 75 gsm rating is

$$\frac{75}{2 \times 2^0} \times 1.05 = 39.38 \text{ kilograms}$$

For the American or British sizes the following formula can be applied:

$$\text{Weight of a ream of paper} = K \frac{\text{gsm rating}}{2} \text{ kgs.}$$

K is a conversion constant and is equal to

Bond = 0.241
Demy = 0.248
Royal = 0.3225
Crown = 0.194

For other sizes $K = 6.4516 \times 10^{-4} \times \text{Length} \times \text{Width}$

(Length and width to be expressed in inches)

Example: The weight of a ream of Demy sized paper in 45 gsm is

$$\frac{0.248 \ (K) \times 45}{2} = 5.58 \text{ kg}$$

9.5.2 Calculating the length of a reel of paper

Weight of the reel = w kgs

Width of the reel = c cm

Substance of paper = g gsm

$$\text{Length of paper in the reel} = \frac{w \times 10^5}{c \times g} \text{ meters}$$

Example: The length of paper in a reel 67 cms wide and weighing 200 kgs is 3732 meters if the paper is 80 gsm.

9.5.3 Calculating the length of a reel of paper (Approximate method)

For a roll of paper let the outer diameter be RD and the core diameter CD. Let it have a width of W centimeters. The weight of the roll is approximated depending on the type of paper used. The weight is calculated as:

Weight of the roll of paper $= F \times (RD^2 - CD^2) \times W$ kg

Here RD and CD are expressed in cm and F is a factor depending on the kind of paper.

F has a value:

For newsprint	0.443×10^{-3}
For Antique finish	0.498×10^{-3}
For MF, offset, bond	0.747×10^{-3}
For SC, one side coated	0.830×10^{-3}
For 2 side coated	0.941×10^{-3}

Example: A roll of offset paper wound on a 10 cm core with an outer diameter of 100 cm and a width of 75 cm will weigh:

$$0.747 \times 10^{-3} \times (10^4 - 10^2) \times 75 = 554.7 \text{ kg}$$

9.6 SPOILAGE ALLOWANCES

9.6.1 Sheetfed work

Quantity	Printer's allowance	Binder's allowance	Total spoilage
1K to 2K	4%	2.5%	6.5%
2K to 5K	3%	2.0%	5.0%
5K to 10K	2.5%	1.5%	4.0%
10K+	2.5%	1.0%	3.5%

Add 2% per colour to printers allowance for multicolour work.

9.6.2 Webfeed work

Quantity	Printer's allowance	Binder's allowance	Total spoilage
10K to 15K	11%	1%	12%
15K to 25K	9%	1%	10%
25K to 75K	7%	1%	8%
75K+	5%	1%	6%

Add 3% per colour for multicolour work.

9.6.3 Lamination

Certain number of prints are required in excess for setting up laminating and UV curing units.

The spoilage may be estimated as follows:

Up to 5K	5% Overs
5K+	3%
100K+	1%

If relative humidity is 50 to 55% and surfaces are supplied flat anything can be laminated with film. When very rough, absorbent and pregrained surfaces exist, a proofing may be necessary. Rough matt coated boards are not suitable for lamination. When UV coating is done the substrate surface properties are even more critical and hence selection should be done more cautiously. (Surface oil absorbency test should have value in excess of 60 seconds is desirable.)

9.6.4 Ink selection for lamination

Warning: Colours may look different after lamination.

The formulation should be such that:

1. They dry quickly with low residual solvent.
2. Have minimal wax, polyethylene and silicone and other surface active agents. They are added to introduce slip which prevents scuffing.

3. Should be tinctorially strong.

4. Pigment should be resistant to solvents as per BS-3442 (Test method 4).

5. UV cured inks are preferred.

6. Metallic inks may get affected by immigration of lubricants to the inks surface preventing adhesion of the lamination. (Avoid laminating metallic ink-coated surfaces.)

7. Poor bonding to metal flakes in metallic inks leading to delamination. (Avoid laminating metallic ink coated surfaces.)

8. If foil stamping is to be done subsequently check.

Appendices

Glossary of Terms Used in Paper Industry & Trade

ANSI: American National Standards Institute, provides standards for all products.

BS: British Standards institution. Specifies standards and conditions for testing graphic products.

Caliper: Thickness of paper expressed either as thousandth of an inch or in points. Normally expressed only with thick papers or boards.

Chemical pulping: The process of separating the fibers from wood free of lignin, by the action of alkali solutions.

Chlorolignin: A water insoluble compound of chlorine and lignin formed during bleaching operations.

Coating: A varnish like substance opaque white or transparent deposited over paper to improve surface and absorption properties.

Cocking: Uneven drying paper causes dimensional errors known as cocking.

Colour management system: A system of representing colour varieties in terms of numbers for colour communication and representation.

Effluent: The outflow of chemical waste from industrial establishments, the environmental problems of which must be taken care of.

Feeding system: The major mechanical component of a sheet fed printing machine. Paper has to move from zero velocity to machine speed and hence has to undergo mechanical pressure. The paper has to stand the pressures of picking up, acceleration and inking in a printing machine.

Fines: Finely cut fiber bundles which easily pass through wire meshes and also the fourdrinier wire.

Flocs: Pulp fibers sticking together in small bundles.

Foils: A system placed under the foudrinier wire to provide a vacuum when the wire runs at high speeds to assist faster draining of water.

Furnish: A mixture of fibers, additives, water and sometimes sizing agents that are mixed for making paper.

GATF: Graphic Arts Technical Foundation, a monitoring authority of progress in printing in US.

Gum Arabic: A water soluble colloid, widely used in paper making as well as printing.

Headbox: The paper machine part that provides the hydraulic head to pulp before distribution to wire.

Hydration: The degree of swelling and softening of fibers during refining.

IGT: Instituut Voor Grafische Technik, Denmark. The institute produces and certifies most of the equipment required for quality control of ink and paper.

Lignin: The binding glue in wood, that need be removed to free the fibers.

nm: Nanometers $= 10^{-9}$ meters.

Printability: The quality of paper to get inked for reproducing text, pictures etc.

Process colour: The system of using 3 or more transparent colours with black to reproduce colour pictures. The colours commonly used are cyan, magenta and yellow with black.

Ream: A packet of 500 sheets of paper.

Section: A book is divided into a set of pages known as sections which are printed on single sheets. The pages are organised or imposed in such a way that when folded the pages are in the correct sequence. A section is the smallest printing size of pages.

Shives: Bundles of fiber formed as large fiber pieces which are light and hence difficult to separate from pulp by centrifugal techniques.

Stock: Same as furnish—a mixture of fibers, additives and water ready to be turned into paper.

Toxicity: Degree of identified poisonous properties of an item.

TAPPI: The American Association for Pulp And Paper. The journal of the association is the standard for paper and pulp industry.

Unwinding tension: The force on paper when it has to unwind form a reel as on a web machine. This is the minimum tension the paper has to stand.

μm: Micrometers, 10^{-6} of meter.

VOC: Volatile organic compounds that are hazardous to the environment. Usually a constituent of ink.

Vol: A measure to indicate the thickness of 100 sheets (200 pages) of paper when bound into a book. 25 vol would mean the sheets would be 25 mm thick for 200 pages. Papers are usually between 14 and 25 vols for book work.

Typical Paper Plant Capacities

Newsprint only	100-200 kTons/yr
Poster papers and writing papers	100 kTons/yr
Boards and writing papers	100 kTons/yr
Soft tissue, unbleached kraft etc.	50 kTons/yr
Bond papers, high brightness papers, duplicating and cream wove papers	10 kTons/yr
Special battery separator paper	100 Tons/yr
Coated papers, PVC coated papers and speciality paper	3-5 kTons/yr

Bibliography

1. **Pulp and Paper**, James P Casey, Editor. Vols. I, II, III.
2. **Printing Paper and Inks**, Charles Finley. Delmar Publications.
3. **The Paper Industry**, Dr. P.E. Sankaranarayanan, Editor, Kothari Desk Book Series.
4. **What the Printer Should Know About Paper?** GATF.
5. **Publishers' Guide to Paper**, PIRA.
6. **Project Profiles of Handmade Paper Industry**, KVIC, Bombay.